S0-ABU-470

Servicing Home Video Cassette Recorders

MARVIN HOBBS

HAYDEN BOOKS
A Division of Howard W. Sams & Company
4300 West 62nd Street
Indianapolis, Indiana 46268 USA

© 1982 by Hayden Books
A Division of Howard W. Sams & Co.

FIRST EDITION
EIGHTH PRINTING—1987

All rights reserved. No part of this book shall be reproduced, stored in a
retrieval system, or transmitted by any means, electronic, mechanical,
photocopying, recording, or otherwise, without written permission from the
publisher. No patent liability is assumed with respect to the use of the
information contained herein. While every precaution has been taken in
the preparation of this book, the publisher assumes no responsibility for
errors or omissions. Neither is any liability assumed for damages resulting
from the use of the information contained herein.

International Standard Book Number: 0-8104-0652-7
Library of Congress Catalog Card Number: 81-7256

Printed in the United States of America

Service personnel should comply with all caution and safety-related notes
located on or inside the VCR or camera cabinet, on their chassis, and in the
manufacturer's manuals and data.

Preface

In the American market it is well known that only two basic formats of home video recorders have been sold in significant quantities: the Beta and VHS formats. Both of them have been sold under a variety of brand names. Although numerous models exist in each format, they differ primarily in the length of record/play time and in programmable features. Therefore, by treating the circuitry and mechanical features of typical recorders with Beta and VHS formats, one can explain most of the home video tape recorders in this country. Two other formats, used in earlier models, are no longer offered, but some of these machines are still to be found in the hands of customers. Also, in Europe and other countries—where European manufacturers have entered the picture—additional formats such as those offered by Philips and Grundig have appeared in recent years. In this text, however, we have usually confined ourselves to video recorders designed to handle NTSC standard TV signals.

Because of basic similarities in the recorders designed for different formats, it is our basic philosophy to describe certain aspects of one format or another in detail and then point out the differences that exist between it and the others. Various chapters cover the video signal circuitry, servo circuit operation, and system control operation of the VHS system in detail. The differences found in Betamax are described separately in yet another chapter. A comparison is also made between the mechanical aspects of these formats. Several chapters treat the adjustment tools and electronic test equipment used for VTR servicing, mechanical adjustments and replacements, and electrical alignment and adjustment. A final chapter covers personal video camera theory and servicing.

Home video tape recorders differ from other consumer electronic products in that practically no electronic test equipment specifically designed to service the electronic and electrical portions of these units has been offered by test equipment suppliers. In the main, NTSC signal generators, dual trace scopes, frequency counters, and digital voltmeters are proposed by most authorities, including those who write the service manuals for the various recorders (of course, with routines specified for the use of these instruments). Only one exception exists—an instrument that was designed to service color TV receivers but where special consideration was given to home video tape recorders and their signal requirements (the Sencore VA48 Video Analyzer). In some measure the lack of specialized electronic test equipment has resulted from the availability of test/ alignment tapes that supply color bar, monoscope, sweep, and audio signals. Such tapes are supplied by recorder manufacturers. Specialized sets of jigs, fixtures, and tools are provided for the adjustment and maintenance of the mechanical portions of home video tape recorders. Differences in the mechanical design of recorders has made it necessary for the VCR manufacturers to supply special tools to enable servicemen to repair their products. Although some of these tools can be used in the servicing of any recorder, there is a tendency for manufacturers of a given format to offer a set of tools meant for their own family of recorders.

Information furnished by home video recorder manufacturers and test equipment companies has made this book possible. The author wishes to thank the Hitachi Sales Corporation of America, Sony Corporation of America, Victor Company of Japan, Radio Corporation of America, and the Zenith Radio Corporation for information on their video recorders and personal video cameras. He also wishes to thank Sencore, Inc., and the Leader Instrument Corporation for information on their electronic test equipment, ETCO Electronics for information on cable TV Up-Converters, and the IEEE and DEMPA Publications for permission to use material from their literature.

MARVIN HOBBS

Contents

Servicing Home Video Cassette Recorders

1

Background for Home Video Recording

As long ago as 1927, Baird in England was experimenting with the recording of low-definition television signals on phonograph records, and another inventor, B. Rtcheouloff, also in England, filed a patent for recording television signals on magnetic material. For at least the next 25 years, recordings for television transmissions, experimental or otherwise, were done on photographic films. Such recordings were made like movies, picked up from the film, and converted to signals suitable for modulating TV transmitters.

From the early 1950s, Ampex in the United States and Toshiba in Japan began development work on the magnetic tape recording of TV signals with approaches that bore some resemblance to present-day methods. However, prior to these efforts the Electronic Division of Bing Crosby Enterprises in California demonstrated a black-and-white video tape recorder in 1951. It used twelve recording heads—ten of which recorded the video signal, one of which recorded a synchronizing track, and the twelfth recorded the audio. By an ingenious method of sampling each head in a stroboscopic manner, an alternating signal with positive and negative halves representing bits of picture information up to about 1.7 MHz was realized from the ten heads. Magnetic tape, with widths ranging from ⅜ to 1 in., was moved at a speed of 100 in./sec.

In 1953, RCA demonstrated a video tape recorder capable of handling color as well as black-and-white signals. A ¼-in. tape, traveling at 360 in./sec., had one track for the black-and-white picture and one for sound. Color recording required five tracks—three for the color signals, one for sync signals, and one for audio. Tape reels 17 in. in diameter made possible recordings of only 4 min in length. Both the Crosby and RCA approaches used fixed magnetic heads and high longitudinal tape speeds. In 1956, RCA

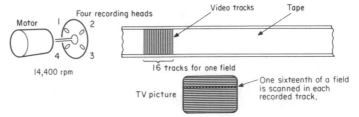

Fig. 1-1 Video recording method used in broadcast equipment (four recording heads with transverse scan on tape)

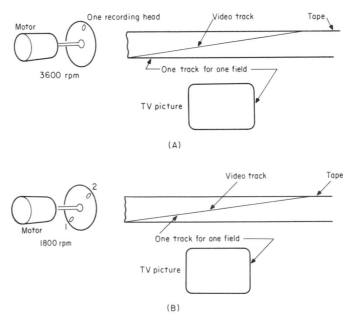

Fig. 1-2 Helical scan (A) with one video head, and (B) with two video heads

reported on another video recorder, in which ½-in. magnetic tape running at a speed of 20 ft/sec (6.1 m/sec) contained seven tracks—three of which were for color, one for mixed high-frequency video components, one for sync signals, and the other two for audio.

Ampex and Toshiba followed totally different approaches. They used longitudinal tape speeds much the same as those used for audio recordings and placed the magnetic heads on cylinders or drums that rotated at high speed relative to the slow-moving tape. In the case of Ampex, four heads were placed 90 deg apart around a cylinder or drum, which rotated at 14,400 rpm and laid down 16 transverse tracks per field perpendicular to the tape's motion (see Fig. 1-1). Toshiba experimented with both single-head and double-head configurations on high-speed rotating drums. However, they recorded at an angle across the tape as it moved along so that one complete picture field was recorded by each pass of a head, as shown in Fig.

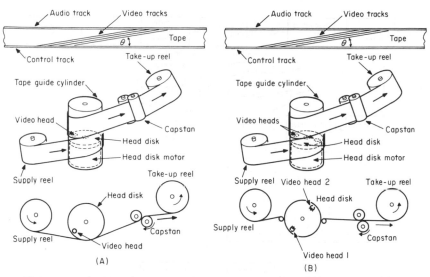

Fig. 1-3 Early Toshiba mechanism (A) with one video head, and (B) with two video heads

1-2. The Ampex approach, which was demonstrated in 1956, was developed more fully toward the end of the 1950s and became known as the Quad Head Transverse System. This system has dominated the broadcast video tape-recording field ever since. The Toshiba approach, which was demonstrated in 1959, was developed more fully over the next decade by both Toshiba and others and became known as the Helical Scan System. This system has penetrated the broadcast TV field in various forms and has become the basis for present-day home video recorder methods.

As a point of history in Japan, N. Sawazaki, the Research Director of Toshiba, is credited with inventing helical-scan video recording in 1954, although his patent was not granted until 1959. Patents were granted to Schuler in Germany in 1955 and to Masterson in the United States in 1956 for similar methods. Nevertheless, Toshiba introduced the first helical-scan recorder for broadcast applications in 1959. Except for the fact that their tape transport was reel-to-reel, this recorder embodied most of the principles that were to be utilized later in home video recorders. In designing their recorder, Toshiba considered the rotary drum with one and two video heads, as shown in Figs. 1-3(a) and 1-3(b), respectively. These were to rotate at a high speed synchronized with the field frequency of the TV picture. Tape was fed at the relatively low speed of 15 in./sec, so that the locus of contact of the recording head with the tape formed evenly spaced slanted tracks on the tape. In the electronic portion of the recorder, Toshiba used the arrangement shown in Fig. 1-4. The video signal was frequency-modulated onto a carrier prior to recording to reduce its octave range (see Chap. 2). However, this recorder did not have a chroma channel, since it handled only black-and-white signals.

But it did employ both drum and capstan servo systems, as shown in Fig. 1-4. Phase detection was accomplished in the head drum servo by a

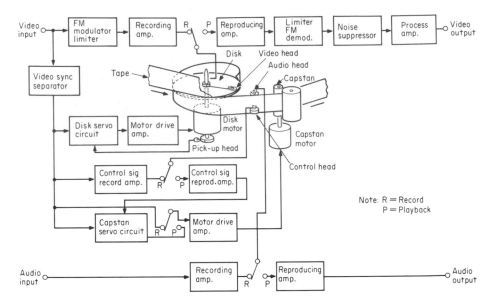

Fig. 1-4 Block diagram of an early Toshiba helical scan recorder

piece of ferromagnetic material placed on its edge. This piece of material induced one pulse in each revolution of the pickup head. The drum motor was driven by the phase-difference signal between the induced pulse and the vertical sync pulse of the TV signal to make the head pass the edge of the tape at an appropriate phase in the blanking interval. In recording with the capstan servo, a vertical pulse was recorded as a control signal along the lower edge of the tape, and the motor was driven by a 60-cycle oscillator. The audio track was recorded along the upper edge of the tape.

The Development of Compact Video Tape Recorders

Neither Ampex or Toshiba used tape cassettes; both used reel-to-reel tape transports. In fact, the first compact video tape recorder introduced in Japan in 1964 was an open-reel type using a ½-in. tape. Two years later, five companies had compact reel-to-reel models using helical scanning on the market in Japan. Although they did not compete in performance with the Quad broadcast models, they sold at about ⅟₁₀₀ of the price of the latter, which was more than $100,000. By 1969, the Electronics Industries Association of Japan sought to standardize the several open-reel designs on the market and announced standards for those recorders using ½-in. tape. Thus, some degree of compatibility between tapes produced by different manufacturers became feasible. However, this class of recorders did not find its way into many homes. Instead, it was used mainly in schools and other enterprises for educational purposes. In this role, it contributed greatly to the progress of audiovisual education.

Various approaches to tape handling for helical-scan video recorders were taken during the 1960s. In general, these ranged from open-reel types to cartridge types with an eye to finding a form that would be easy to handle and relatively foolproof. Although the complexity and costs of such designs relegated them for the most part to professional applications, cartridge types were developed in some cases with an eye to possible home use. Since none of these designs had any impact on the home market, we will begin our review of the further development of home video recorders with cassette types. Interest in the video field began to accelerate about the same time that cassettes started to replace open-reel tapes in the audio field. Two of the earliest developers in the application of cassettes to the video-recording field were Sony and Philips.

In April 1969, the Sony Corporation of America announced their first color cassette video tape recorder. It used a 1-in. tape running at 3½ in./sec with two tracks. The cassette, measuring about 6 × 10 × 3¾ in. (152 × 254 × 95 mm), had the supply and take-up reels mounted coaxially. The tape came off one reel, passed around a hub that was slightly larger than the head drum, and wound onto the second reel. When the cassette was slipped into the recorder, the hub slipped over the drum, locking the cassette in place and rotating the hub so that a window in it let the tape contact the drum. The luminance signal was recorded by the frequency modulation of a subcarrier, whose frequency was about 3–4.5 MHz. The NTSC color signal was shifted below the FM modulation spectrum to 900 kHz and recorded with a bandwidth of ±700 kHz around the center frequency.

In November 1969, Sony announced another home recorder using a ¾-in. tape running at 3.15 in./sec enclosed in a cassette measuring 8 × 5 × 1¼ in. (203 × 127 × 32 mm). This unit, designed to have a record or playback time of up to 90 min, was a forerunner to the U-Matic system, which appeared in 1971. Signal processing was based on NTSC standards, and a 250-line resolution was claimed. The previously mentioned recorders as well as the U-Matic design were the pioneers of the home video recorders available today. An outline of the cassette used in the U-Matic system is shown in Fig. 1-5.

In the U-Matic design, which was actually standardized by Sony, JVC, and Matsushita, the signal processing shown in Fig. 1-6 took place. The luminance signal is recorded as an FM carrier frequency, which is 5.4 MHz for peak-white and 3.8 MHz for sync tip. The subcarrier of 3.58 MHz which carries the chrominance signals, is converted to 688.37 kHz. Before recording, this down-converted subcarrier is mixed with the FM luminance signal. Linearity of this subcarrier is maintained in recording because the FM luminance signal, being at a higher frequency and having a greater amplitude, acts as a bias. The mixing ratio between the FM luminance signal and the down-converted subcarrier is critical. Lower chroma subcarrier level causes noise in the chrominance signal, whereas a higher chroma subcarrier level causes beating and generates spurious signals. Color stability is achieved by an automatic phase control circuit for the reproduced subcarrier and a stable tape transport mechanism.

In the playback mode of the U-Matic system, the reproduced signal from the tape is filtered to separate the luminance and chrominance components. The amplitude of the chrominance component is controlled by an AGC circuit so that the burst level is correct regardless of the

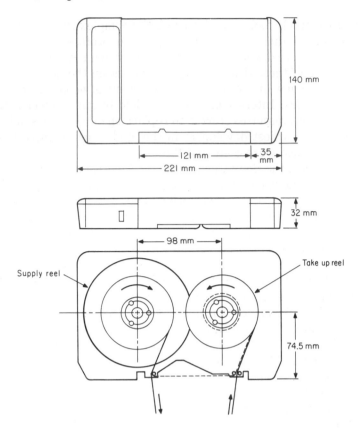

Fig. 1-5 U-matic professional video cassette

fluctuation of the reproduction level. The reproduced burst signal, which is gated, controls the phase of the local oscillator that reconverts the subcarrier frequency to 3.58 MHz. In this way, jitter components are eliminated at each horizontal scan sequence. The FM luminance signal goes through conventional reproduction circuits, equalizing amplifiers, limiter, and FM demodulator, and is mixed with the reconverted chroma subcarrier. The output signal then takes its original NTSC form. The reason we discuss the U-Matic recorder in so much detail is easily explained. Although it is a professional machine and not much used in the home, its basic circuitry and design is so much like that of present-day home recorders that it can be considered to be a true parent in their family tree.

As mentioned previously, Philips also was one of the early developers of cassette-type professional recorders. In June 1970, they announced details of a deck measuring 22 × 13 × 6½ in. with cassettes measuring 5. 8 × 5.0 × 1.4 in., for both monochrome and color. In this design, the cassettes were coaxially mounted reels of chromium dioxide (CrO_2) tape, which ran at 5.6 in./sec. Their system also incorporated sound dubbing, independent recording of the audio tracks, and *stop motion*. Here again Philips's work in professional recorders helped pioneer their home units using ½-in. tape, which they began to introduce in 1972.

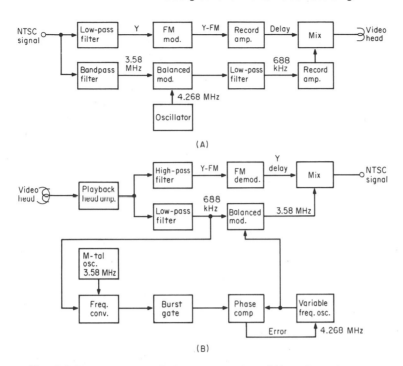

Fig. 1-6 Block diagram of electronic portion of U-matic professional video recorder: (A) record mode, and (B) playback mode

Home Video Development

As the major Japanese color TV manufacturers began to see that market heading for a sales plateau, they looked for another video product to offset it. Although it was still a relatively high-priced consumer item, they chose the home video tape recorder (VTR) to fill this role. Not expecting it to reach the sales volume of less-expensive products immediately, they considered it to have at least three main sales features. One was its ability to record programs off the air. Second was its ability to play back prerecorded tapes available from sources other than the TV receiver. Third was the fact that color video cameras can be used in conjunction with VTRs to supplement home movies. However, to meet consumer requirements, the manufacturers saw that long-running recording and playback was necessary along with compactness and a tape package that was easy to handle. In the early 1970s, there were limitations to the achievement of ultimate goals in these directions.

In 1974, Toshiba and Sanyo Electric introduced VTRs designated as V-Cord I and II. These machines were unique in that they employed a skip-field system to give them longer recording/playing modes of 2 hr (a long operating time at that date). This was done by skipping every other field of the picture. Thus, every other TV picture was recorded, and the same frame was played back twice to give an impression of continuity. These recorders offered 1 hour of recording and playback in a standard mode and 2 hours in the skip mode.

In 1975, Sony opened a new chapter in the history of video cassette recording by announcing their Betamax, which featured *azimuth recording*. Although their design had a basic 1-hr record/play capability with guard bands between the tracks, they could extend the playing time to 2 hr by overlapping the adjacent tracks and thereby eliminating the guard bands. Such an overlap would have resulted in cross talk without a special technique of adjusting the angles of the head gaps relative to each other. When the gaps were set at a certain angle from the perpendicular in opposite directions, the crosstalk between adjacent tracks could be greatly attenuated through *azimuth loss* (see Chap. 2).

Despite Sony's advance, Matsushita introduced a single-head machine, the VX-2000 (sold in the United States as a Quasar brand model VR1000 and called the "Great Time Machine"), and Toshiba and Sanyo introduced the V-Cord II with three heads. Then, in the autumn of 1976, JVC (Victor Company of Japan) introduced its VHS system with two heads, which provided a basic 2-hr record/playback capability. By 1977, rivalry between the Beta format and the VHS format groups began in earnest. Sony, Toshiba, and Sanyo formed the Beta format group in Japan, while Matsushita, JVC, and Hitachi formed the VHS group. All of these companies undertook the production of video recorders to supply U.S. outlets, selling under various brand names. In the United States Beta format recorders were sold by Sony, Zenith, Toshiba, Sears Roebuck, Sanyo, Pioneer, and Aiwa. VHS recorders were sold by RCA, General Electric, Panasonic, JVC, Magnavox, Montgomery-Ward, Sylvania, Hitachi, Curtis Mathes, MGA, J.C. Penney, Sharp, and Akai. However, all of these recorders were made in Japan by manufacturers in the Betamax or VHS group

Fig. 1-7 Typical VHS video recorder with two- and four-hour modes (*Courtesy,* Victor Company of Japan, Ltd.)

mentioned previously. A typical VHS recorder of that period with a 2-hr record/play capability is shown in Fig. 1–7. Models were introduced later with 2/4/6-hr capability.

In the meantime, both Betamax and VHS recorders, meeting PAL and SECAM color standards wherever necessary, were introduced to the European market as well as the rest of the world. The only competition that the Japanese VTR manufacturers have faced has been that provided by Philips and Grundig, especially in Europe. Until 1979, both these companies offered designs that used the coaxial cassettes that they had adopted several years earlier. While these gave excellent picture quality, it was debatable whether this type of cassette could command a substantial market in the face of worldwide distribution of recorders using Beta and VHS cassettes and the availability of wide selections of prerecorded tapes in these formats.

However, in 1979 Philips and Grundig introduced their Video 2000 system with a unique cassette, which, after operating in one direction for 4 hr, can be turned over like an audio cassette to provide 8 hr of record/play time in a single package. Not only is this system, in models meeting PAL and SECAM standards, sold in Europe and the world market, but it also may be offered in models meeting NTSC standards in the U.S. market (perhaps, even in the Japanese market).

As of 1981, no U.S. manufacturers had undertaken the production of a video tape recorder for home use.

2

The Basic Elements of Home Video Recorders

Despite differences in the formats of home video tape recorders, a number of basic aspects of their electrical and mechanical design are similar. These are as follows:

1. All video home tape recorders depend on narrow-gap, high-performance magnetic heads for the recording and playback functions.
2. All depend on high-speed transit of the tape relative to the magnetic head or vice versa. Either rotating heads must pass by the tape at high speed, or tape must pass by one or more fixed heads at a rapid rate.
3. All depend on the use of frequency modulation of the luminance signal onto a carrier prior to recording to reduce the octave range between the lowest and highest frequency to be recorded.
4. All use the *color-under* method to place the chroma signal frequency below the FM band of the luminance signal prior to recording.
5. All use some form of servo system to keep the recorded and playback signals in synchronism with the television signal being processed.
6. All use tape cassettes and have automatic means for associating the tape with the recording/playback heads after the cassette has been entered into the recorder.
7. All depend on magnetic tape, which has been improved in many ways since the introduction of quadruplex video tape recorders to broadcasters in the late 1950s.

8. All use azimuth recording to permit the overlapping of adjacent recorded tracks (sometimes called *overwrite recording*) in long play modes to extend the recording time on a given length of tape.

Video Head Development

When one considers the vast difference between a frequency range of 50 to 20,000 Hz for the recording of audio signals and a frequency range of 0 to 4.2 MHz for the recording of video signals, it is quite apparent that magnetic heads for the recording and playback of video signals must be far superior to those used in audio recorders—at least, in their ability to record high-frequency signals. The video head has been described as the heart of a video tape recorder.

From the following formulas, it may be seen that the gap width of the video head and the relative speed of the head and the tape are key factors in realizing the wideband frequency capabilities essential to video recording.

First is the relationship between relative speed of the head and the tape and the recorded wavelength on the tape, as expressed by formula (1):

(1) $$\lambda = \frac{V}{f}$$

where

λ = recorded wavelength (cm/cycle)
V = tape speed (cm/sec)
f = input signal frequency (cycles/sec, Hz).

Second, from the characteristics of magnetic recording playback, the relationship between the minimum recorded wavelength and the gap of the video head can be expressed by formula (2):

(2) $$g = \frac{\lambda}{2}$$

where

g = gap width of video head
λ = required minimum recorded wavelength on the tape

From formulas (1) and (2) it is apparent that the gap of the video head must be made narrower in order to decrease the relative speed of the head and tape and expand the frequencies to be handled into a higher range.

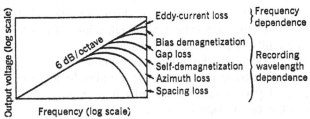

Fig. 2-1 Factors affecting frequency response in magnetic recording (H. Sugaya, Matsushita; Transactions on Magnetics ©, IEEE)

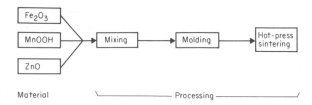

Fig. 2-2 Materials and processing used in making a video recording head

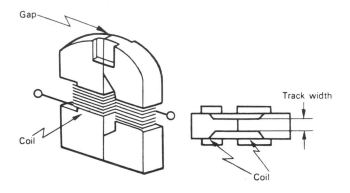

Fig. 2-3 Typical video recording and playback head

Conventional magnetic heads have output following the 6-dB/octave law, which is shown in Fig. 2-1. However, their high-frequency performance is influenced by several factors, some of which have frequency dependence and several of which have recorded wavelength dependence. Even though the step from audio heads to video heads for broadcast applications as initially used by Ampex in their quadruplex recorders represented a quantum jump, the home video application required a video head of even higher efficiency and more durability to increase the recording density. Ferrite appeared to be an ideal material to meet these requirements, but the brittleness of conventional ferrite caused by its porosity required the development of a better form of this material. Two different types of ferrite resulted. One of these is called *single-crystal ferrite* and the other, *hot-pressed ferrite*. The latter was developed mainly by Matsushita Electric in Japan and carried their trademark (HPF). When the head track width is reduced, as shown in Fig. 2-5, the grain size of the hot-pressed ferrite becomes comparable to the head track width. In order to solve this problem, Matsushita developed a crystal-oriented hot-pressed ferrite material in 1977. Although the service person will never need to become involved in manufacturing the ferrite for a video head, Fig. 2-2 indicates the materials and processes that go into it and may give him a better appreciation of the technology involved.

A sketch of a video head is shown in Fig. 2-3. As will be described in detail later, typical home video recorders contain two such heads mounted on a high-speed rotating drum to provide the necessary speed of the head

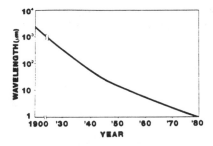

Fig. 2-4 Progress in reducing the recording wavelength in video head development (H. Sugaya, Matsushita; Transactions on Magnetics ©, IEEE)

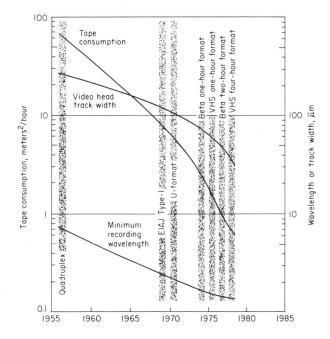

Fig. 2-5 Progress in reducing tape consumption, video head track width, and minimum recording wavelength (H. Sugaya, Matsushita; Transactions on Magnetics ©, IEEE)

relative to the tape. Figures 2-4 and 2-5 show the progress that has been made in reducing the recorded wavelength and the video head track width, particularly since the inception of video recording in the late 1950s. In present-day home video heads, the ferrite material is about 3 mm square and about 100 microns in thickness. The coils of the head are made of wire finer in diameter than a human hair. The head gap has been narrowed to give the smallest track width possible.

High-Speed Transit of Tape Relative to Magnetic Head (or Vice Versa)

While progress in video head development has gone a long way toward reducing the speed required to provide the necessary wide band

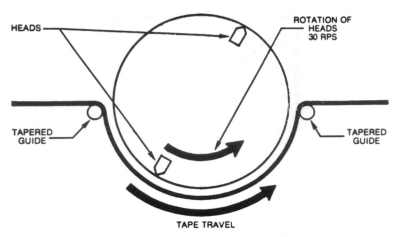

Fig. 2-6 Rotating drum carries the video heads at high speed relative to the recording tape (*Courtesy,* RCA)

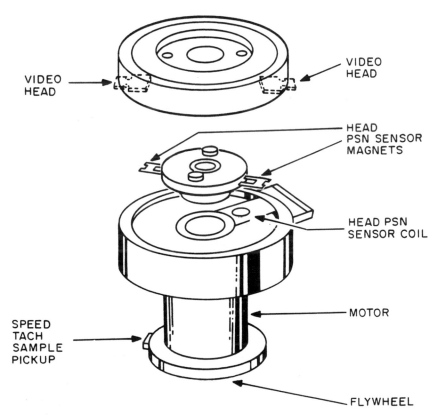

Fig. 2-7 Exploded view showing video heads, head sensor magnets, drive motor, and speed tach pickup (*Courtesy,* RCA)

recording, it is still necessary to use writing speeds far in excess of those required for audio recording. In most home video recorders, the necessary speed relative to the tape is achieved by placing the heads on a drum, as shown in Fig. 2-6, which rotates at a speed of 1,800 rpm (30 rps) for NTSC television standards and at 1,500 rpm for PAL and SECAM standards (used in many countries outside the United States). This high speed of the heads as they rotate past the tape permits the tape to travel at speeds comparable with those of audio recorders. Typically, such speeds are 1.3 in./sec for standard play and 0.65 in./sec for long play operation for the VHS format and a little faster for the Beta format. The manner in which the video heads are attached to the drum is shown in Fig. 2-7. The tape is pulled past the rotating heads by a capstan drive working against a pinch roller much as audio tape is pulled across the fixed head in such systems. Figures 2-8 and 2-9 show top and bottom views of a typical VHS recorder, indicating the location of the head drum, the capstan, and their drives.

Feeding around guides adjacent to the rotating drum, the video tape also is pulled past a combination audio and control track head on one side and past an erase head on the opposite side of the drum. The capstan may be driven by a separate motor, as shown in Fig. 2-10, which is the arrangement found in VHS recorders; or it may be driven by the same ac motor that drives the rotating drum, as shown in Fig. 2-11. The latter arrangement is found in Beta format models.

Since 1956, when the rotating head method was first introduced in quadruplex video recorders for broadcast use, the rotating head has dominated the field. As a result, helical-scan recording has been developed as the method for home VTRs in contrast with the quadruplex method for

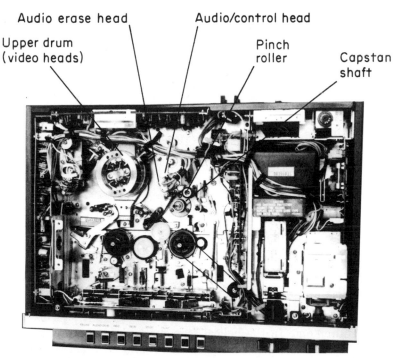

Fig. 2-8 Top view of chassis of a typical VHS recorder (*Courtesy,* Victor Company of Japan, Ltd.)

Capstan flywheel   Capstan
                                    motor

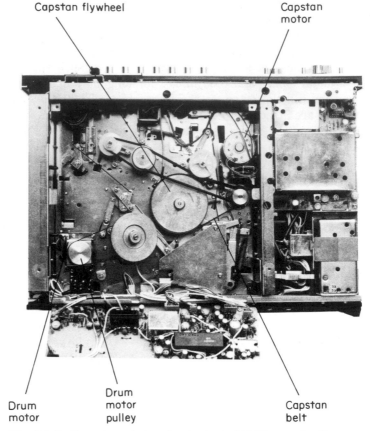

Drum   Drum   Capstan
motor   motor   belt
              pulley

Fig. 2-9 Bottom view of chassis of a typical VHS recorder (*Courtesy,*
Victor Company of Japan, Ltd.)

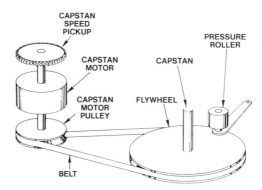

Fig. 2-10 Capstan drive by separate motor (*Courtesy,* RCA)

broadcast equipment. In the helical-scan method, each head records one
field on each pass, as shown in Fig. 2-12 (A) and (B). The audio signal is
recorded along one edge of the tape, and the control track is recorded along
the opposite edge.

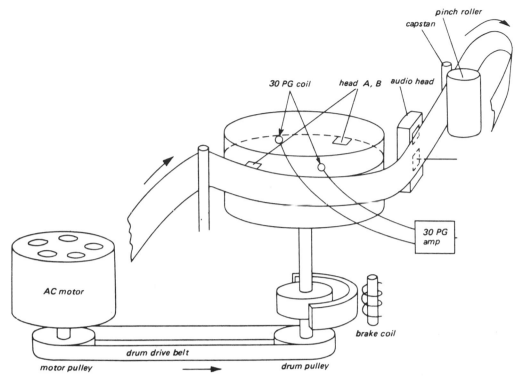

Fig. 2-11 Drum (disk) drive by separate motor (*Courtesy,* Sony Corp. of America)

Longitudinal recording, in which the tape traveled past a fixed head at a high speed, was tried initially more than 20 years ago in experimental video recording before helical-scan recording was developed. At that time, video head development had not progressed to the point where such a method was feasible. However, development came at such a pace that by 1978 Toshiba was able to demonstrate a fixed-head recorder with a 1-hr playing time, which they could extend to 2-hr operation. In this model (see Fig. 2-13), the tape traveled at 6 m/sec and required 220 tracks laid down along its width to provide the record/play time indicated above in a reasonable physical space.

The Use of Frequency Modulation

The problem of recording and reproducing the high frequencies of video signals has been solved by narrowing the gap of magnetic recording heads—improving their high-frequency response by their construction— and by revolving them to increase the relative speed of the tape and the head. However, these steps still are not sufficient for the recording of the very wide video frequency band that extends from at least 30 Hz to more than 4 MHz. The remaining problem arises from the fact that the playback performance of video heads is almost completely proportional to frequency,

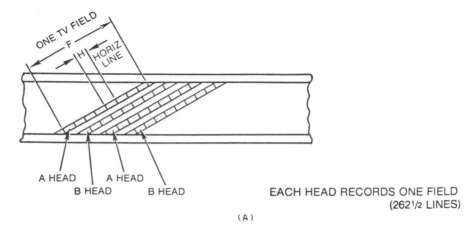

EACH HEAD RECORDS ONE FIELD
(262½ LINES)

(A)

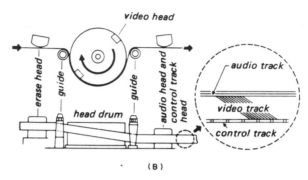

(B)

Fig. 2-12 (A) Pattern of recorded tracks in a dual-head recording, and (B) components involved in producing video, control, and audio tracks (*Courtesy,* Sony Corp. of America)

Recording pattern of fixed head system

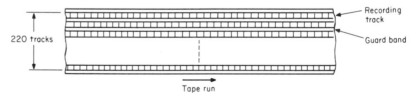

Fig. 2-13 Recording pattern of Toshiba's longitudinal fixed-head video recorder

especially within their linear output range. As mentioned previously, this relationship is known as the 6-dB/octave law (see Fig. 2-1). In short, regardless of whatever output can be realized at 4 MHz, the output falls off at a rate of 6 dB/octave as the frequency decreases. Since the video signal (ranging from 4 MHz on the high side to 30 Hz on the low side) has a range of more than 18 octaves, it is easy to see that the low-frequency portion of the signal could not be recorded or played back without some steps to reduce the octave range.

The octave range of the video signal can be reduced by mixing it with any frequency higher than the lowest video frequency and taking the sum of the frequencies resulting from the mixing. Mixing with a low frequency would not accomplish much, but mixing with 4 MHz, say, would reduce the octave range to a single octave, as shown in Fig. 2-14(A). Amplitude modulation could also be used to reduce the octave range. Let us say a carrier of 5 MHz is amplitude-modulated by the video signal. In this case, the lower sidebands would extend to 4 MHz below the carrier, or to 1 MHz, and the upper sidebands would extend to 4 MHz above the carrier, or to 9 MHz. If the upper sidebands are eliminated by filters and we use only the lower sidebands for recording purposes, signals ranging from 1 to 5 MHz will be available. The octave range is now reduced to between two and three octaves, as seen in Fig. 2-14(B).

While mixing or modulating the video signal onto a higher frequency solves the problem of reducing its octave range for recording purposes, it is better to frequency-modulate this carrier rather than to amplitude-modulate it. FM is preferred because it is less sensitive to amplitude changes, which occur as a result of irregularities in the recording process. Also, the use of FM permits one to set the sync-tip and peak-white levels to specific frequencies so that in playback these levels will be restored and give accurate replicas of the input signal.

To solve this problem, the luminance portion of the video signal is used to frequency modulate a carrier prior to recording. In a typical case, a carrier frequency of 3.4 MHz would be frequency-modulated with sync tips appearing at 3.4 MHz and peak-whites at 4.4 MHz, as shown in Fig. 2-15. It is seen that the lower sidebands of the frequency-modulated signal extend to about 1.2 MHz. Since only the lower sideband needs to be reproduced, it is seen that the octave range of the signal to be recorded has been reduced to a modest figure by using a frequency-modulated carrier for the recording.

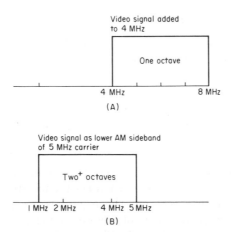

Fig. 2-14 (A) Reduction of octave range by mixing with a fixed carrier, and (B) reduction of octave range by amplitude modulation of a 5-MHz carrier

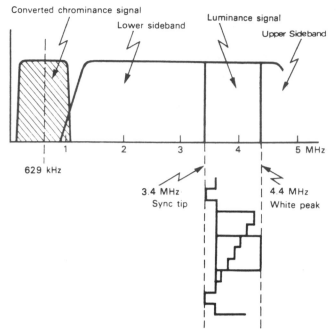

Fig. 2-15 Using frequency modulation of the luminance signal to reduce the octave range

Upon playback, the FM signal from the video heads can be passed through limiters and other circuits where irregularities in output due to either the heads or the tape can be compensated for or eliminated. Also, the need for ac bias, which is required in conventional audio recording, is eliminated. Once the FM signal has been fully processed, it is demodulated and the luminance signal is mixed back with the chroma signal to provide a full color signal with original characteristics.

The Color-Under Technique

Since the interleaving principle is used for the multiplex transmission of the chrominance signal (consisting of hue and chroma-saturation-information), it cannot be recorded along with the FM luminance signal without modification. A solution has been to down-convert the chrominance signal subcarrier, which is at 3.58 MHz in an NTSC standard TV signal, to a frequency below the lower FM sideband. In the case of the Beta format, this lower frequency is approximately 688 kHz, and in the VHS format, it is approximately 629 kHz.

As shown in Fig. 2-16(A), the frequency-modulated luminance signal follows one path and the down-converted chrominance signal follows another path. Prior to recording, these signals are mixed together, but they are still separated in terms of frequency prior to being recorded, as shown in the frequency spectrum in Fig. 2-16(B).

The advantages of converting the chrominance signal to a lower frequency are as follows:

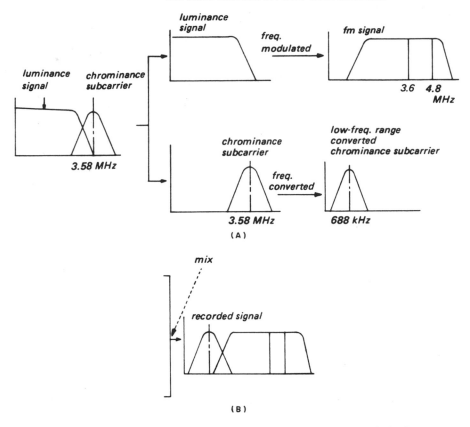

Fig. 2-16 Processing of the luminance and chrominance signals in the "color under" technique as applied in Betamax recorders (*Courtesy, Sony Corp. of America*)

1. No bias is required to record the chrominance signal. The FM modulated luminance signal, which is recorded together with the chrominance signal, accomplishes the same result as a bias current.
2. Electronic stability is good at the lower frequency.
3. It is easy to include a circuit to remove the effects of jitter.

The Need for Servo Control

It has been shown previously that the development trend has been to make the recorded tracks on the tape smaller and smaller. Track widths of less than 0.03 mm are not uncommon. No matter how precisely the tracks are recorded, a good picture cannot be reproduced if these tracks are not traced accurately during playback. Making all the parts in the tape path as mechanically precise as possible will not suffice. In addition, all home VTRs must employ a self-governing arrangement—known as a *servo system*—to maintain accurate tracing and picture stability.

During recording, the vertical sync signal within the video signal is synchronized with the rotating heads and governed by a pulse signal (30

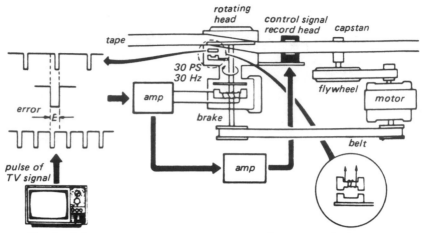

Fig. 2-17 Servo action in the record mode of Betamax unit (*Courtesy,* Sony Corp. of America)

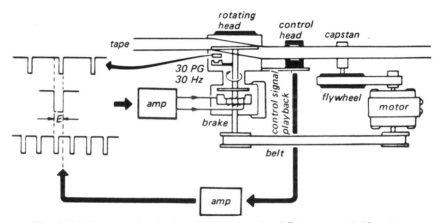

Fig. 2-18 Servo action in the playback mode of Betamax unit (*Courtesy,* Sony Corp. of America)

Hz) that is generated in such a way that the heads rotate at precisely 1800 rpm (30 rps) for the NTSC standard. In a typical design in which a single ac motor drives the head drum and the capstan, as shown in Fig. 2-17, the servo action is accomplished as follows:

1. The heads are driven by a belt so that they turn just a little faster than 1,800 rpm (30 rps).
2. A 30-Hz signal is derived from the vertical sync signal (60 Hz) by dividing its frequency. This is used as a control signal (CTL signal) and recorded on the tape by a separate stationary control head.
3. A pulse signal (30 PG) generated by detecting the actual rotational speed of the heads is compared with the vertical sync signal (30 Hz). Any difference (error signal) between the two is amplified

and the appropriate amount of current is applied to the brake coil to correct the deviation instantly. In this way, rotational speed is maintained at precisely 1,800 rpm. In other words, the current applied to the brake coil is increased if rotational speed exceeds 1,800 rpm and decreased if the speed drops below that figure.

During playback, as shown in Fig. 2-18, the control signal recorded on the tape becomes the standard reference signal. The control track head picks up this control signal, which is compared with the 30-PG signal of the rotating heads. By synchronizing these two signals, both rotational speed and phase are made the same as they were at the time of recording. In this way, playback picture stability is achieved. This particular design is referred to by Sony as their *drum servo*.

Tape Cassettes

The first home video recorders used reel-to-reel tape transports. While this method was accepted for audio tape recorders for many years, because high-quality cassettes were not available, it was never considered to be the answer for home video tape recorders. It was never felt that the average user could be expected to thread video recording tape around a rotating drum and past all of the other points where it had to travel. In the late 1960s, there was considerable debate about whether video tape for home recorders should be housed in cartridges or cassettes. Cartridges were much more popular than cassettes for home audio tape players at that time. However, Sony Corporation and the Victor Company of Japan introduced video cassette recorders in 1969, and Philips followed in 1970. Cassettes finally won the debate because they are easier to handle in recording operations. Cartridges were acceptable for audio players, primarily because recording was not the main consideration. When recording became a bigger factor, especially relative to the use of audio players in cars, the cassette dominated the audio scene, too.

In general, video tape cassettes took two forms. One was the coplanar type, in which the supply reel and the take-up reel are side by side, a form that has since been most widely used. Such a form is seen in the cassettes of Sony and JVC (Victor Company of Japan) in their Betamax and VHS systems, respectively. Although they differ in dimensions, as shown in Fig. 2-19, and also differ in the recording format on the tape that they contain, they are of the same general family. The other type of cassette is the coaxial or stacked type, in which the supply reel and the take-up are on the same axis; that is, one is stacked over the other. This form of cassette, shown in Fig. 2-20, has been called the "Philips" type, and indeed it has been used by Philips and their affiliate—Grundig—in all of their home cassette video recorders introduced prior to the Philips System 2000. The stacked cassette also was used initially in some designs by JVC and was later marketed in the United States by Matsushita in their model VR1000 recorder, called the "Great Time Machine."

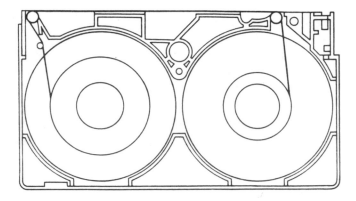

Video tape	Length	T-120	248m ± 1m	(120 min.)
		T-60	128m ± 1m	(60 min.)
		T-30	68m ± 1m	(30 min.)
	Width	12.65 ± 0.01 mm		
	Thickness	0.02 mm		
Leader tape and trailer tape	Length	170 + 5 mm		
Dimensions		104(D) × 188(W) × 25(T) mm		

		Betamax X2	Betamax
A	tape width	12.65 mm (½ inch)	12.65 mm (½ inch)
B	video track pitch	29.2 μm	58.5 μm
C	video track width	10.62 mm	10.62 mm
D	control track width	0.6 mm	0.6 mm
E	audio track width	1.05 mm	1.05 mm

Fig. 2-19 A comparison of some VHS and Betamax cassettes

In their System 2000, Philips uses a unique reversible coplanar cassette that is capable of handling at least 8 hours of video recording and playback by recording on only one-half of the tape width in each direction.

With the tape housed in a cassette and placed in the recorder, it must be brought into proximity with the magnetic heads housed in the rotating drum. This action is accomplished automatically inside the recorder, when the Record or Play buttons are pushed. Figure 2-21 shows basically what the loading operation must accomplish. The position of the tape when the cassette is first entered into the recorder is shown by the solid lines between the supply reel and the take-up reel. All recorders use some form of loading mechanism to bring the tape out of the cassette and cause it to take the path around the drum. One path is shown by dotted lines in Fig. 2-19. Either a motor associated with the capstan drive or a separate motor operates the loading mechanism. When the tape takes this path, its

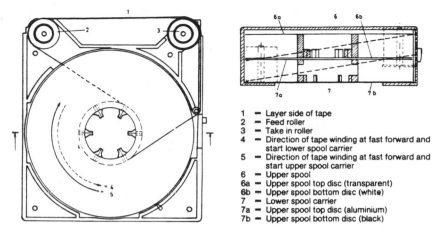

1 — Layer side of tape
2 — Feed roller
3 — Take in roller
4 — Direction of tape winding at fast forward and start lower spool carrier
5 — Direction of tape winding at fast forward and start upper spool carrier
6 — Upper spool
6a — Upper spool top disc (transparent)
6b — Upper spool bottom disc (white)
7 — Lower spool carrier
7a — Upper spool top disc (aluminium)
7b — Upper spool bottom disc (black)

Fig. 2-20 Coaxial or stacked type of cassette ("Philips" type)

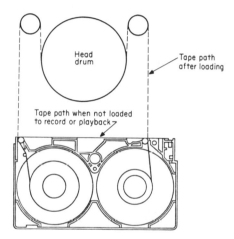

Fig. 2-21 Basic loading operation of video tape from the cassette to the rotating drum as in VHS recorders

automatic handling is called *M-loading*, because of the shape of the path taken. M-loading is used in VHS recorders. In Betamax recorders, Sony uses another type of loading called *U-loading*, which again derives its name from the configuration of the tape path—the letter *U* on its side.

The Importance of Magnetic Tape Development

Video tape, as well as the video head, has an important role in increasing the recording density, which determines the maximum recording time attainable within the limits of acceptable picture quality. When the quadruplex recorders were developed for broadcast use in the late 1950s, the magnetic tape used for video recording had almost the same performance characteristics as that used for audio recording at the time. Figure 2-22 shows a physical comparison of present-day audio and video tapes. In order to increase the recording density of the tape, many factors

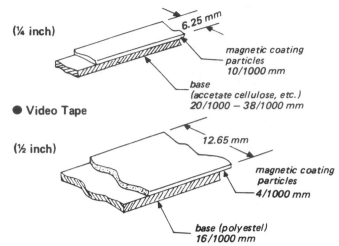

Fig. 2-22 Physical comparison of video and audio tape (*Courtesy,* Sony Corp. of America)

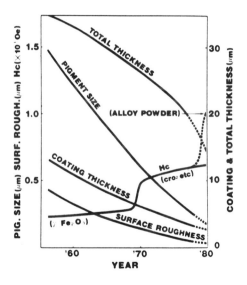

Fig. 2-23 Improvement of video tape during two decades (H. Sugaya, Matsushita; *Courtesy,* IEEE)

had to be improved. The extent to which they have been improved over the past 20 years is shown in Fig. 2-23.

Important characteristics to consider in recording tape are the magnetic flux in the tape and the magnetic force required to produce it. The former is designated by the symbol B and the latter by the symbol H. The relationship between these two characteristics for a typical magnetic material (in this case, the recording tape) is expressed by the B-H curve, shown in Fig. 2-24.

In practice, magnetic materials retain a certain amount of the magnetization or magnetic flux produced by a magnetizing force even after

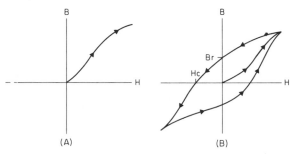

Fig. 2-24 (A) B-H curve showing magnetic flux produced by the magnetizing force, and (B) hysteresis curve indication the coercivity, Hc

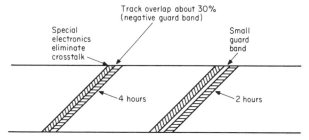

Fig. 2-25 Comparison of overlapped recorded tracks with tracks separated by guard band

that force is reduced to zero. Thus, if the magnetizing force (H) is increased to produce a given magnetic flux (B), it is not possible to move along the ideal curve of Fig. 2-24(A) if one reduces the magnetizing force after reaching a certain level. Instead, the curves of Fig. 2-24(B) result to give the so-called *hysteresis loop*. In this loop, the points B_r and H_c are important in that they indicate, respectively, the amount of magnetization that remains in the magnetic material when the magnetizing force is reduced to zero and the amount of magnetization force that must be applied to reduce the magnetization to zero. The former is called the *remanence*, and the latter is called the *coercive force* or *coercivity*. It has been shown that the high frequency sensitivity of tape depends on the coercivity (H_c) and that the medium and low frequency sensitivities depend on the remanence (B_r). Thus, the frequency response is proportional to the ratio of the two: H_r/B_r. Coercivity is often expressed as follows: $0_e = 10^3 \times H_e$.

The *coercivity* (H_c) of the tape was gradually increased in order to decrease the self-demagnetization factor. By 1970 the coercivity had been doubled in so-called *high-energy tape*, such as CrO_2 (chromium dioxide) tape. Such tape was adopted for the U-format professional recorders at that time. In 1973, two Japanese tape manufacturers—Fuji Film and TDK—independently developed Co-modified iron-oxide tape that simplified the production process. Since that time, either CrO_2- or Co-modified iron-oxide tape has been used for home VTRs. However, progress in tape improvement did not stop there.

The pigment size of the magnetic material has been decreased considerably over the years to reduce the noise component and to increase the coercivity of the tape. Use of alloy powder will continue to decrease the pigment size in the future. Smoothness (or the opposite of roughness) of the tape surface has its good and bad aspects. The tape surface should be as smooth as possible to decrease the nonmagnetic space between the head and the tape, and as shown in Fig. 2-23, it has become smoother over the years. However, the smoother the surface of the tape, the more difficult it is to pass it around the head drum, and *jitter* can be caused in the playback picture. Even worse, a smoother surface may shorten the life of the heads because of the close contact between the head and the tape as a result of high tape tension. So there must be a practical compromise in tape smoothness.

Tape thickness also has been decreased considerably over the years, leading to increased recording time for a given cassette size. However, thinner tape presents its problems by leading to difficulty of tape movement, lack of machine-to-machine interchangeability, and unfavorable skew characteristics. Moreover, since the stiffness of the tape is related to the cube of its thickness, the only solution is a compromise. Compromise is literally the theme of video tape manufacturers and the most difficult aspect of production, because the binder system must always satisfy all the conditions listed below:

1. Higher output and signal-to-noise ratio, especially at short wave-lengths
2. Easy and perfect erasability
3. Lower head wear
4. Better tape movement
5. Smaller skew characteristics
6. Longer playing time
7. Machine-to-machine interchangeability

Coercivity affects items 1 and 2 in the previous list; pigment size affects items 1 through 3; total thickness affects items 3 through 6; surface roughness is significant in items 1, 3, and 5; and the binder system is a factor in items 3 through 5.

Since magnetic tape is the lifeblood of the video tape recorder, it should always be ahead of magnetic recording technology. A further look at Fig. 2-23 shows that the slopes of the total thickness curve and the coercivity curve show the greatest change in recent years. Their improvement has led to longer playing times, higher output, and better signal-to-noise ratios as well as lower head wear. The greatest promise for higher density recording in the future may lie in the further development of magnetic alloy powders for the magnetic part of the tape.

The Elimination of Guard Bands and the Use of Azimuth Recording

Early models of home video recorders used guard bands between the tracks on the tape, as shown in Fig. 2-25. This method separated the tracks

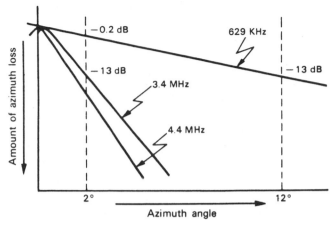

Fig. 2-26 High frequency loss due to azimuth angle of playback head relative to recorded track (*Courtesy,* Hitachi)

well enough for crosstalk not to be a major factor. Since the home video recorder has been sold in greater quantities, the demand for increased recording and playing time has continually risen. To extend the playing time of a given recorder and cassette design in a long mode, manufacturers have slowed the rate of tape travel and overlapped the adjacent tracks on the tape. Obviously, crosstalk will result when tracks are overlapped. Therefore, all home video recorders employ a method (called *azimuth recording*) to reduce the effects of crosstalk to an acceptable level.

In Japan, the initial development of the azimuth method is credited to Professor Okamura of the University of Electro-Communication, who patented it in 1959. Of course, Sony, JVC, Matsushita, and others have contributed to its development and application since that time. Azimuth recording takes advantage of what is known as *azimuth loss*, which is a distinct disadvantage in conventional audio recording. For the latter, the higher the frequency, the higher the signal loss that will arise as a result of the difference in the angle of the recording head and the playback head. In video recording with the azimuth method, the angles of the heads recording adjacent tracks A and B are different. In playback, if head A traces track A, the output will be normal, but if head A runs over track B, the B signal will be so weak—because the gap angle of head A is different—that it will barely cause any interference.

Consequently, the azimuth method slants the heads at an angle, θ, instead of positioning them so that their gaps are at right angles to their forward motion direction. Head A is slanted forward by the angle θ, and head B is slanted backward by the same angle. When the video signals are recorded alternately by these heads, two tracks with different magnetization directions are recorded alternately in accordance with the slanted angles of the head gaps. When played back, the head used in recording traces the tracks, and signals are picked up. However, because the magnetization direction of the adjacent track on the tape is different, the azimuth loss of the head increases, high frequencies are attenuated, and crosstalk is minimal.

In Betamax recorders, the angle θ is ± 7 deg; for VHS recorders, angle θ is ± 6 deg; and for the Philips System 2000, angle θ is ± 15 deg. However, they are all using the same basic method just described.

Despite the effectiveness of azimuth recording in correcting for the crosstalk at higher frequencies, it should not be forgotten that the method is not comparably effective in dealing with lower frequencies. Since the color carrier is down-converted to a frequency below the frequency-modulated video band, it is at a lower frequency than the luminance signal in video recording. The variation of azimuth loss with azimuth angle at video frequencies is shown in Fig. 2-26. The azimuth system as just described is not effective in handling chroma crosstalk between adjacent tracks when they are overlapped. Thus, we find another system added in the chroma channels of video recorders to deal with chroma crosstalk. These methods, which vary between the different systems, are described in detail in Chaps. 3 and 4.

3

The Formats of Home and Personal Portable Video Recording

Since the introduction of the technique of rotating magnetic heads at high speeds in a head drum to obtain the high writing speeds required to record video signals, there has been a trend toward reducing the width of the magnetic tape. Video head and tape development and helical scanning have led to a reduction from a tape width of 2 in. in early broadcast recorders to as low as 1/4 in. in the newest formats. The latter is found both in the compact video cassette of Technicolor's personal portable recorder made by Funai in Japan and in the 8-hr cassette of the Philips System 2000 home recorder. In the Philips recorder, two ¼-in. tracks are laid down in opposite directions on a ½-in. tape. Of course, neither of these are compatible with the widely used VHS and Beta formats using the ½-in. tape width. Nevertheless, the trend to narrower tape continues, and more and more recorders will make use of it. This development will especially take place in small personal portable recorders combined with video cameras. In addition, the likely introduction of longitudinal-scan home and/ or portable recorders will lead to still further incompatibility.

Helical-Scan ½-in. Tape Formats

Betamax and VHS Compared

It may often appear to the shopper that the only differences in various models of Betamax and VHS recorders are in the record/play time, the dimensions of the cassettes, and some of the ancillary features. However, there are also more internal differences, such as their color-signal process-ing systems and their tape-loading mechanisms, as well as differences in head-drum diameters and tape-writing speeds. In so far as the latter are concerned, the head-drum diameter in the Betamax system is 74.49 mm; in the VHS system, it is 62 mm. In Betamax recorders the longitudinal tape speed is 40 mm/sec, and the relative writing speed on the tape, taking into account the rotating heads, is 6.9 m/sec. In VHS systems, the longitudinal

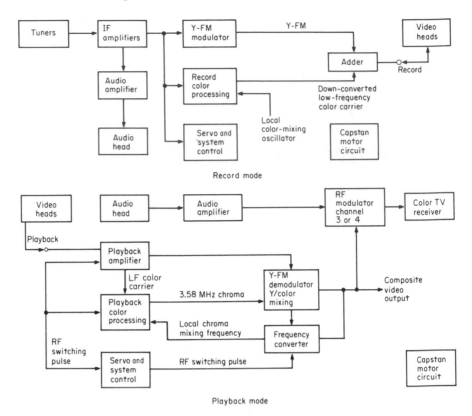

Fig. 3-1 Basic block diagram of VHS and Betamax recorders: (A) Record mode, and (B) Playback mode

tape speed is 33.4 mm/sec, and the relative writing speed on the tape is 5.8 m/sec. These tape speeds are for standard play modes.

The lower tape speed and the smaller drum of the VHS system makes it possible to fit enough 20-mm-thick tape into a cassette, measuring only 188 × 104 × 25 mm, to obtain 2 hours of record/play time. A disadvantage of the smaller drum in VHS systems is a reduction in the relative speed between the video heads and the tape. It is slowed to 5.8 m/sec compared to 6.9 m/sec in Betamax systems. Such slowness decreases the maximum recordable wavelength of the highest FM frequencies, a limitation that tightens the requirements on the heads, tape, and circuits.

The Betamax cassette is smaller than the VHS cassette, measuring only 156 × 96 × 25 mm. It is this aspect of the design that is responsible for much of the operating difference of the two systems. The original Betamax system was actually designed for 1 hour of record/play time, whereas the VHS system was designed for 2 hours of record/play time. The internal VHS design dimensions, regardless of whatever problems they may have presented to their manufacturers, made it possible to obtain this longer playing time in a 2-hr cassette that was only slightly larger than the original 1-hr Betamax cassette. From then on, as market demands for longer record/play time steadily increased, the Betamax systems were multiplied up from a 1-hr base, whereas VHS systems were multiplied up for a 2-hr base. Of course, both systems were multiplied by 2 to give 2 hr and 4 hr by reducing linear tape speeds to one-half of their standard record/play rate. Further extensions of time have been provided by reducing tape thickness. Betamax systems have been pushed up to 4 and 4½ hr of record/play time in newer designs, and VHS systems have been pushed up to 6 hr maximum.

In general, Betamax and VHS recorders follow the basic block diagrams shown in Fig. 3-1. However, major differences are found in their internal designs. One difference is the way in which the chroma portion of color TV signals are handled to eliminate crosstalk between overlapping recorded tracks; another is the way in which tape is loaded relative to the rotating head drum after the cassette has been placed in the recorder. Both of these design differences will now be discussed.

A major difference in the circuitry of Betamax and VHS systems results from the color-processing methods used for both the record and playback modes. Both systems use the *color-under* method and record the color signal without going through the frequency modulation channel employed by both systems for the luminance signal. A difference in the color-under signal is that in Betamax it is converted to approximately 688 kHz, whereas in the VHS system it is converted to approximately 629 kHz. However, most of the differences in the two systems in this regard arise from their methods of eliminating chroma crosstalk from track to track on the tape. In Betamax, Sony uses a phase-inversion recording method to eliminate crosstalk between chroma signals on adjacent tracks by interleaving their frequency spectrums. This method divides alternate tracks on the tape into A and B groups and processes them in a manner that rejects interference between them. In the VHS method for eliminating crosstalk between adjacent tracks, the phase of the chroma signal along a given track is shifted 90 deg for each successive horizontal line in the field. These methods will now be described in greater detail.

The Beta Phase Inverter Color Recording System

In Beta systems, the chroma signal on one track (called *A*) has its phase inverted 180 deg with every line period, while the chroma signal on

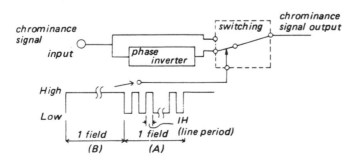

Fig. 3-2 Record mode switching for beta color method (*Courtesy,* Sony Corp. of America)

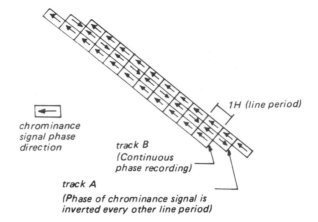

chrominance
signal phase
direction

track B
(Continuous
phase recording)

1H (line period)

track A
(Phase of chrominance signal is
inverted every other line period)

Fig. 3-3 Recording phase pattern in beta color system (*Courtesy*, Sony
Corp. of America)

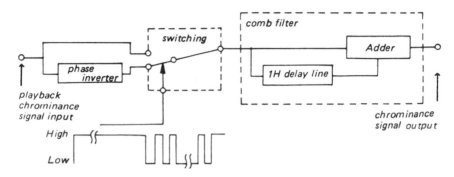

Fig. 3-4 Playback mode switching for beta color method (*Courtesy*, Sony
Corp. of America)

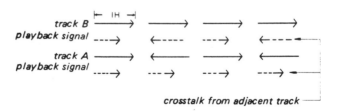

Fig. 3-5 Crosstalk phase relationships in beta playback (*Courtesy*, Sony
Corp. of America)

the other track (called *B*) remains continuously in the same phase. In the
record mode, the electronic switch, as shown in Fig. 3-2, goes up with a
high-level signal and down with a low-level signal. Therefore, within the A-
field period the signal will change phase by 180 deg on track A every line
period and be recorded in the pattern shown in Fig. 3-3. In the playback
mode, the electronic switch, which is shown in Fig. 3-4, goes up with high-
level and down with low-level input, and the crosstalk will appear as shown
in Fig. 3-5. Since the track-A playback signal has its phase inverted with

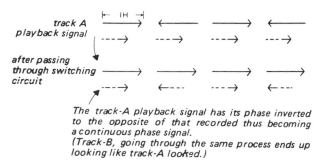

The track-A playback signal has its phase inverted to the opposite of that recorded thus becoming a continuous phase signal.
(Track-B, going through the same process ends up looking like track-A looked.)

Fig. 3-6 Processing to obtain continuous phase signals in playback (*Courtesy*, Sony Corp. of America)

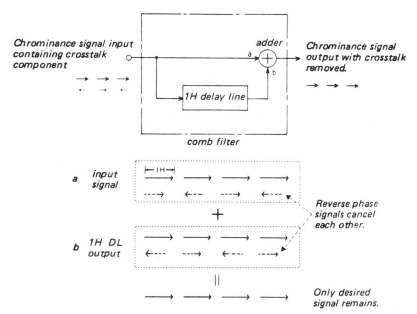

Fig. 3-7 Elimination of chroma signal crosstalk by comb filter method (*Courtesy*, Sony Corp. of America)

every other line period (every field or every 1 H), it is converted back into a normal continuous phase signal, as shown in Fig. 3-6.

As a result of these interactions, the playback chrominance signals and the crosstalk components of both tracks A and B come to have the same phase relationship. By passing the signal through a comb filter using a 1-H delay line, the crosstalk component is eliminated, and only the desired signal emerges, as shown in Fig. 3-7. The complete playback process is shown in Fig. 3-8.

Because of the employment of phase-inversion color recording, it is necessary to convert the signal back to its original form during playback. Automatic Phase Control is important in accomplishing this result. Hence, the down-converted chroma signal is locked to the horizontal sync component in the video input signal by an AFC and APC system. Figure 3-9 shows a block diagram of the chroma signal system.

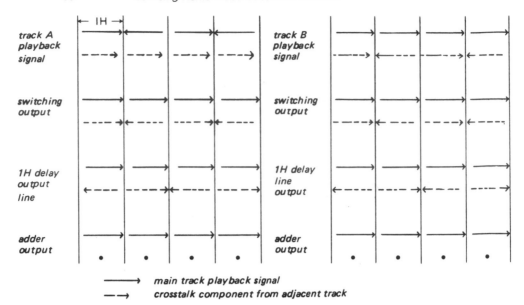

→ main track playback signal
--→ crosstalk component from adjacent track

Fig. 3-8 Phase relationships on adjacent tracts in beta playback process (*Courtesy,* Sony Corp. of America)

The Beta Chroma System in the Record Mode

The 3.58-MHz chroma signal is applied to a frequency converter. A carrier signal of 3.58 MHz + (44 − ¼) fH = 4.27 MHz also is applied. The converter mixes the chroma signal down to (44 − ¼) fH = 688 kHz.

An AFC loop, which is a PLL (phase-locked loop) circuit, produces a 44-fH frequency from the H-sync signal separated from the input video signal in the record mode (fH indicates the H-sync frequency). The output of the 44-fH VCO (Voltage-Controlled Oscillator) is divided in a divide-by-44 countdown circuit. The output signal from the countdown circuit is phase-compared against the H-sync signal in a phase comparator that yields a dc output voltage. This dc output is applied to the VCO as the control voltage. The output frequency of the VCO is exactly 44 fH locked to fH, the H-sync frequency of the incoming video signal. The output of the crystal oscillator, 3.58 MHz − ¼ fH, is applied to frequency converter circuit II, and the output of that converter is 4.27 MHz = 3.58 + (44 − ¼) fH, the sum frequency of the the two signals. The crystal oscillator acts as a fixed-frequency oscillator in the record mode and as a variable-frequency oscillator in the playback mode.

The 4.27-MHz output is applied to frequency converter I on the record side. As already mentioned, there the signal is converted to 688 kHz after being derived from (44 − ¼) fH and is thus locked to fH, the horizontal sync frequency. This 4.27-MHz signal is supplied to the frequency converter via a phase inverter circuit, which inverts the phase of the carrier signal, as mentioned previously. The switch, which is shown directly above frequency converter II in Fig. 3-9, is driven by the output of a flip-flop. The flip-flop is triggered by the 30-PG signal and the H sync.

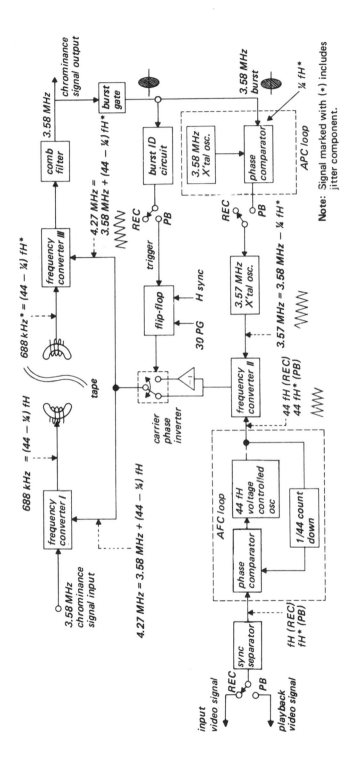

Fig. 3-9 Block diagram of the beta chroma signal system (*Courtesy,* Sony Corp. of America)

This switch is switched every 1 H in the A field and set to the "1" position in the B field. In this form the down-converted 688-kHz chroma signal is recorded on the tape along with the FM, which contains the luminance information.

The Beta Chroma System in the Playback Mode

The 688-kHz chroma signal after being separated from the balance of the reproduced output from the video heads is converted to 3.58 MHz in frequency converter III. In that converter it is mixed with the 4.27-MHz carrier signal applied from the same circuit used in the record mode. The input signal to the AFC loop in the playback mode is the H-sync signal separated from the demodulated video signal. The 3.57 MHz − ¼ fH crystal oscillator acts as a variable-frequency crystal oscillator controlled by the APC loop output.

The frequency of the recorded chroma signal is (44 − ¼) fH locked to fH, the frequency of the horizontal sync signal. When the signal recorded in this way is played back, the relationship between the phase variation of the H-sync signal in the demodulated video signal and the one of the demodulated chroma signal is expressed by the formula; $X = (44 - ¼) Y$, where Y is the phase variation of the H-sync signal of the reproduced video signal and X is the phase variation of the demodulated chroma signal. The 44-fH output, locked to the reproduced H sync, is obtained from the AFC loop; it contains the phase-variation component equal to the one of the chroma signal. Since this signal is used as the carrier signal for converting the chroma signal to 3.58 MHz, the phase-variation component of the chroma signal is canceled in the process of frequency conversion and becomes the 3.58-MHz chroma signal with stabilized phase.

The phase variation of the chroma signal can be eliminated only by the AFC loop, where the H-sync signal is used as a pilot signal. Since the frequency of the down-converted chroma signal is (44 − ¼) fH and the output frequency of the AFC loop is 44 fH, the AFC loop is used for the elimination of the phase variation of ¼ fH. The phase of the burst portion of the chroma signal converted to 3.58 MHz and the one of the 3.58-MHz crystal oscillator output are compared. The resultant error voltage is used for controlling the 3.58 MHz − ¼ fH variable-frequency crystal oscillator.

The phase stabilization of the reproduced chroma signal is attained as just described. The phase of the reproduced chroma signal is shifted every 1 H in the A-field period. It is necessary to restore the phases to the original continuous phases. For this purpose, the phase of the frequency conversion carrier is inverted every 1 H in the A-field period, as it is processed in the record mode. In the playback mode as in the record mode, the flip-flop is triggered by the H-sync signal so that the output of the same switch is responsible for carrier phase switching in either the record or playback modes. When the switched phase becomes opposite to the one in the record mode, the chroma signal is inverted 180 deg in both the A-track and B-track fields. The probability that the switch phase becomes opposite is 50 percent because it is determined by the state of the flip-flop. However, it is prevented from inverting the chroma signal in both tracks by the burst ID (identification) circuit shown above the APC loop in Fig. 3-9. The 3.58-MHz crystal oscillator output is phase-compared with the burst signal. The ID circuit detects that the phase of the burst signal inverts 180 deg at the transition of the fields and supplies a trigger pulse to the flip-flop so as to restore the switch phase to the normal state. Crosstalk in the playback chroma signal from the adjacent tracks is greatly reduced by means of the comb filter that follows frequency converter III.

The action of the comb filter has been described only briefly. In more detail, it is a delay line (to delay the signal by one horizontal line) and a resistor bridge. The bridge has three input points and two output points. The first input is for the nondelayed signal and the others for the delayed signal. One of the output terminals will provide a signal only if the delayed and nondelayed signal are both of the same phase, and the other output terminal will provide an output if the two signals are 180 deg out of phase. This second output provides the desired chroma signal without the crosstalk. The reason that the two signals are separated is that the delayed signal takes a different path around the resistive bridge than that taken by the nondelayed signal, as shown in Fig. 3-10. The left-hand side of the resistor bridge adds the two signals if they are in the same phase. The signals are canceled if they are out of phase. In the right-hand side of the bridge the delayed signal flows through the resistors in the opposite direction and is inverted 180 deg. Now, the signals will add only when they are of opposite polarities. The effect of the comb filter in removing crosstalk from the chroma signal has already been illustrated in Fig. 3-7.

The VHS Color Phase-Shift Recording System

In the VHS system, the crosstalk coming from adjacent tracks is removed from the color signal by a phase-shift method. During recording, after down-conversion to 629 kHz, a chrominance signal with the phase advanced 90 deg every 1 H (field) is placed on channel-1 track, and a chrominance signal with the phase delayed 90 deg every 1 H (field) is placed on channel-2 track. Such a recording system can be expressed vectorially on the tape pattern, as shown in Fig. 3-11. During playback, the signal phase of each successive horizontal line is rotated 90 deg in the direction opposite that in which it was recorded to restore it to the same phase as in the original signal. This successive phase shift during recording and

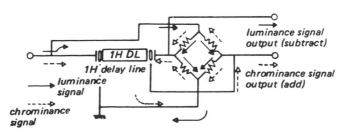

Fig. 3-10 Basic comb filter operation (*Courtesy*, Sencore, Inc.)

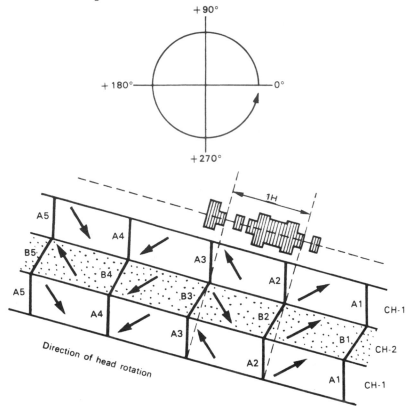

Fig. 3-11 VHS color phase shift recording method (*Courtesy,* Hitachi)

playback permits the desired rejection of the crosstalk that results from the elimination of guard bands in the recording.

In Fig. 3-11, the signal components are designated A and the crosstalk components are designated B. When the channel-1 track is played back by the channel-1 head, A1 and B1 will be played back with the phase delayed by 90 deg every 1 H. The role of a delay line in removing the B or crosstalk component is shown in Fig. 3-12. The relationship of the signal component and the crosstalk component can be looked at as follows:

Signal component: $A_o + A_1$, $A_1 + A_2$, $A_2 + A_3$, etc.
Crosstalk component: $B_o + B_1$, $B_1 + B_2$, $B_2 + B_3$, etc.

where $A_n = A_n + 1$ and $B_n = B_n + 1$ because each of the color television scanning lines can be considered to have essentially the same signal as the adjacent line.

Based on the assumption that the color signal is the same as that of an adjoining area 1 H or one field apart, the signal component adds to become twice as large and the crosstalk component cancels.

The video signal circuitry of a complete VHS recorder (the Hitachi Model 5000A) is described in Chap. 4. The reader is referred to that section for a detailed description of the VHS chroma signal processing circuitry so that he may compare it with the detailed description of the Betamax chroma signal circuitry just covered.

Another difference in the treatment of the chroma signal in Betamax and VHS systems is that in VHS recorders the color burst signal is emphasized by 6 dB in the record mode and deemphasized by a corresponding amount in the playback mode. This procedure, which provides a 6-dB improvement in burst signal-to-noise ratio, is said to be desirable by VHS designers because noise in the color burst translates into phase jitter in the color-processing circuits. Therefore, such emphasis and de-emphasis improve color convergence. Further details are given in Chap. 4.

Tape Loading

A major difference in the Betamax and VHS systems lies in the method of loading the tape from the cassette around the rotating drum that contains the video heads. In Betamax, a method known as *U-loading* is employed, whereas in VHS, a method known as *M-loading* is used. Both methods derive their names from the respective shapes of the tape's path around the drum. As shown in Fig. 3-13, the tape path in the VHS method is seen clearly to form an M-shape. In the case of Betamax, the tape path is U-shaped even though the U is lying on its side (Fig. 3-14).

When a cassette is inserted first into a recorder and its cover is closed, all tape is inside the cassette. In the VHS system, a separate tape-loading motor is provided to operate the mechanism that pulls the tape from the cassette and wraps it around the drum. In Betamax recorders, the same motor that operates the drum and the capstan operates the tape-loading mechanism and causes the U-loading ring to bring the tape into the proper path. Unloading operations are the reverse of loading operations in both systems. These place the tape back into the cassette before it is to be removed from the recorder.

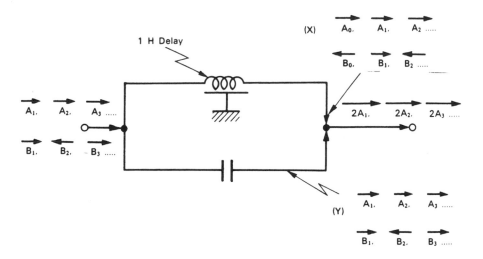

Fig. 3-12 Function of the 1H delay line in VHS chroma system (*Courtesy,* Hitachi)

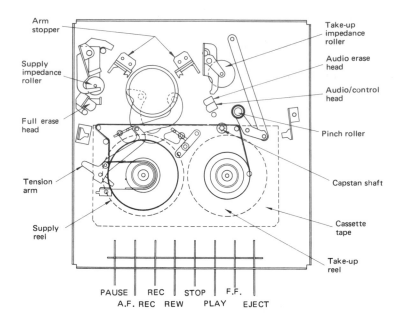

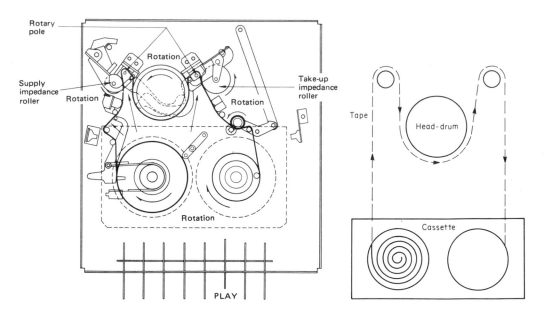

Fig. 3-13 Loading of tape from casette to head drum position in VHS systems (*Courtesy, Journal of the Electronics Industry,* Japan)

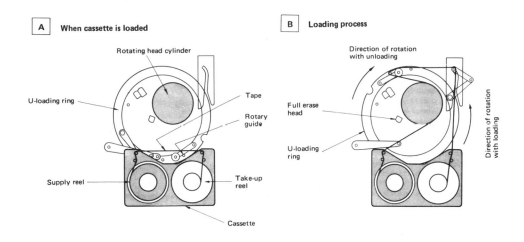

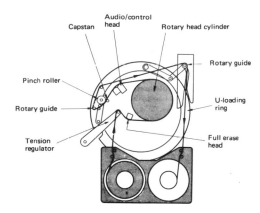

Fig. 3-14 Loading of tape from cassette to head drum position in Betamax systems (*Courtesy, Journal of the Electronics Industry,* Japan)

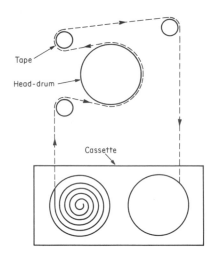

Fig. 3-14 Loading of tape in Betamax systems (cont'd)

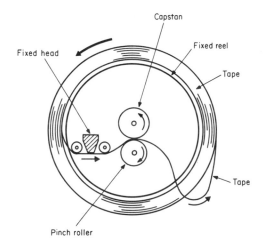

Fig. 3-15 Tape path in Toshiba's longitudinal video recorder

Helical-Scan ¼-in. Tape Formats

The first video recorder using ¼-in. tape to appear on the U.S. market was Technicolor's Model 212. This type of unit appeared initially in Japan as Funai's Model F-812V, which featured a very compact cassette measuring 106 mm in width, 12 mm in height, and 68 mm in depth. The cassette contained 60 meters of tape and was designed to provide 30 minutes of recording time.

The F-812V employed a two-head helical-scan design with a longitudinal speed of 32.1 mm/sec and a video track pitch of 25 microns. The head cylinder diameter was 60 mm, providing a relative tape speed of 5.1 m/sec (almost the same as that of VHS recorders, which is 5.8 m/sec). The azimuth angle was ±11 deg. The audio track pitch, initially 1 mm, was to be reduced in later models to 0.35 mm.

The complete portable VTR measured 246 mm in width, 76.5 mm in height, and 259 mm in depth, with a weight of 3.2 kg when equipped with batteries. It was definitely a step in the direction of a design suitable for use with a video camera in a fully portable lightweight unit.

The Philips System 2000 home recorder, with a somewhat similar recording track, has not yet reached the U.S. market. Its objective was just the opposite of the Funai portable in that maximum recording time was sought. This recorder uses a reversible ½-in. tape in a coaxial cassette. A ¼-in. track providing 4 hours of time is recorded in one direction, and another ¼-in. track is recorded in the opposite direction to provide another 4 hours of time after the cassette has been turned over like an audio cassette. In contrast, VHS and Beta cassettes are not reversible and cannot be turned over. Thus, the objective of the Philips unit is to crowd the maximum amount of recording time in a given space, whereas the objective of the Funai (Technicolor) unit is to provide an acceptable amount of recording time for a given application in the smallest possible space.

Longitudinal-Scan Formats

Although longitudinal-scan, fixed-head video tape recorders were unsuccessful for more than two decades after the development of transverse and helical-scan systems in the late 1950s, developments in 1979 indicated that they may be a factor either in the home or portable field in the future. The reduction of the magnetic head gap and the development of high density tapes have raised the possibility that relatively simple mechanisms in fixed-head machines could bring about the necessary reductions in their size and cost.

At the Consumer Electronics Show in Chicago in June, 1979, Toshiba of Japan unveiled a compact fixed-head recorder measuring 250 × 140 × 330 mm and weighing 8 kg. However, Toshiba made no promises of its availability as a home or consumer portable model. Instead, they seemed to favor industrial and special applications, such as video files, data memories, and product demonstration and monitoring.

In this unit, called Model LVR, a cartridge contains 100 meters of ½-in. tape wound around a fixed reel. The tape travels past the head at a speed of 6 m/sec, which means that in a single pass only 17 seconds of record/play time will be available. To make possible a record/play time of 1 hr for the 100-ft tape in the cartridge, the magnetic head shifts up or down across the tape as the latter passes by so that 220 tracks are recorded on the ½-in. tape in 60 min. This movement of the head gives easy random access to the 17-sec segments of the recording all along the tape throughout the 1-hr span.

As shown in Fig. 3-15, when the cartridge is loaded into the VTR and descends, the fixed head, which is sandwiched between guide posts, is inserted inside the reel. The tape is then drawn out from the window inside the reel by the pinch roller and capstan located near the center of the reel. It makes immediate contact with the head, starts to travel, and begins to record or play back, depending upon the mode into which it is switched. The tape then changes vertical direction with the tape guide provided inside the cartridge. It is twisted once and jumps over the reel; then it is led onto the outside of the reel and is wound up on its outer circumference. The reverse side of the tape has a graphite coating to ensure smooth tape travel.

The width of the recorded tracks is 50 microns, and there are 10 micron guard bands between the tracks. The luminance signal is frequency-modulated to 3.9 MHz at sync signal peak and to 5.4 MHz at white peak. After down-conversion to a sub carrier of 688 kHz, the chrominance signal is added to the FM signal to be recorded. In these respects, the circuitry is seen to be similar to that of helical-scan recorders. Toshiba claims that the horizontal resolution of the picture is better than 240 lines and that the signal-to-noise ratio is better than 42 dB.

Variations Within the Formats

If one counts all the brand names on the U.S. market, more than 100 models of home video tape recorders have been offered to the public. Since many of them are essentially duplicates of Japanese units, the exact number of distinctly different designs is somewhat less. About 25 percent of them are in Beta format and the remainder in VHS. This difference reflects the fact that the great majority of U.S. and Japanese companies adopted the VHS format.

Despite the large number of models, all VHS units operate on the same basic record and playback principles, as do the Beta units. However, different design features appear in many models within a given format for the following reasons:

1. The quest for the longest possible playing time with a given size of cassette
2. The introduction of fast-scan and freeze-frame modes
3. The desire to program recording operations over extended periods of time with optimum flexibility
4. The incorporation of electronic push-button tuning of the TV receiver section of the recorder
5. The desirability of remote control of as many features as possible

Extended Record/Playback Time

As mentioned previously, the VHS cassette originally designed for 2 hours of record/playback time lent itself better to extensions of operating time. Consequently, single cassettes are now capable of up to 6 hours of operation. A number of VHS models providing for 2, 4, and 6 hours of operation have been introduced. Most of the design effort to accomplish these extensions has centered around the crowding of greater lengths of thinner magnetic tape into standard cassettes and running the recorder at slower rates of speed.

If the recorded track width is optimum for shorter recording times, however, it is too wide for optimum performance for longer playing times because of the crowding of the tracks in the play mode. To overcome this drawback in extended play recorders, one design incorporates two pairs of

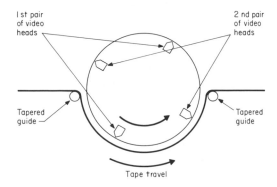

Fig. 3-16 Typical use by some VHS recorders of two pairs of video heads on the rotating drum to provide better performance over a wide range of tape speeds

video heads placed around the rotating drum, as shown in Fig. 3-16. In this particular design, only 2- and 6-hr modes are provided. The video heads used for the 2-hr mode have a track width of 58 microns, whereas the pair of heads for the 6-hr mode have tracks about one-third as wide.

In another three-speed design, which incorporates still- or freeze-framing, Panasonic uses two pairs of heads with dissimilar track widths within each pair. The 2-hr heads are 70 and 90 microns wide; the slow-speed heads are 26 and 31 microns wide. The asymmetrical tracks offset the noise bars in the still-frame operation, placing them outside the visible portions of the frame. The use of offset methods to eliminate noise bands is discussed briefly under microprocessor control of variable speed playback on page 49. For a given record/playback time, a wider head gap is beneficial in effecting the offset, but it obviously has disadvantages as well. For all this, there are VHS models on the market that operate in 2- 4-, and 6-hr modes with only one pair of heads.

The original design of the Beta format provided a 1-hr mode on their L-500 cassette and a 90-min mode on their L-750 cassette. With half-speed operation, 2 hr could be obtained on the L-500 cassette and 3 hr on the L-750 cassette. To compete in the race for longer playing time, a third slower speed was introduced to provide 4½ hr on an L-750 cassette and 5 hr on a new L-830 cassette.

In both VHS and Beta formats, slower speeds result in a loss of video frequencies at the high end, but the picture quality is still considered sufficiently acceptable to warrant the longer record/playback times. Some of the loss in the higher video frequencies is compensated for by means of pre-emphasis and de-emphasis techniques.

The Use of Microprocessors

Microprocessors are used in home VCRs in at least three different ways:

1. For controlling a programmable timer, which permits recording of several different programs and channels over an extended period

of time. The microprocessor turns the recorder on and off and controls the changing of channels in accordance with a program designed to select the desired channels over a period of time.

2. For controlling the functions of the recorder electronically rather than mechanically or electromechanically.

3. For controlling fast and slow scan operation to eliminate certain undesirable effects resulting from playing back a recording at a different speed from that at which it was recorded. Some recorders on the market actually display noise bands in these modes even when operating normally because nothing has been incorporated in their design to eliminate such effects.

Function Control

As shown in Fig. 3-17, a microprocessor may be used to control the operating functions of either a portable or home recorder. In this case, it controls the following functions:

1. *Control of mechanisms and their sequence of operations*: The microprocessor processes inputs from seven operation switches (audio dubbing, record, rewind, play, fast forward, stop, and

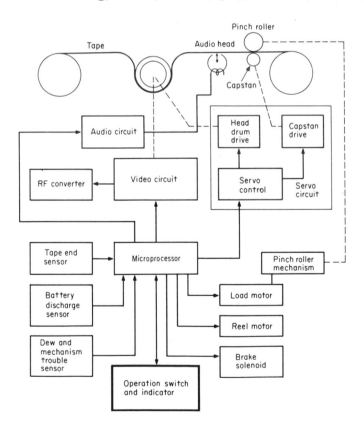

Fig. 3-17 Microprocessor control of the operating functions (M. Ozawa et al, Hitachi, Ltd., Consumer Electronics, © IEEE)

pause), lights indicators, judges whether operations are possible, controls the mechanism, and so forth. Previously, sequence control was exercised mechanically with links, lock plates, and the like. With the use of a microprocessor, this control is fully electronic.

2. *Control over loading and unloading*: Control over the tape loading and unloading mechanism is carried out in response to sensors located in the mechanism.

3. *Control when malfunctions occur*: LED indications are given and necessary operations not damaging to the tape are performed according to inputs that indicate whether condensation has occurred, the motor is locked, there is mechanical trouble, the battery is discharged, and so forth.

4. *Control of automatic assembly recording*: Automatic Assembly Recording, which has application in portable VCRs, is described in Appendix A. Under the control of the microprocessor, the tape is rewound a short way during Record/Pause. When this mode is released, this rewound section of tape is used to match the head phase before recording starts so as to eliminate distortion or scrambling at the junction between each recording when the Pause control is operated. This operation, called the *matching phase* (MP) *system*, is described in Appendix A.

5. *Tape end detection control*: This control function detects the start and end of the tape with sensors and stops the mechanism automatically.

6. *Control of other auxiliary functions*: In a portable VCR when Record/Pause or still Play are continued for more than a specified time, the microprocessor control automatically changes over to a Power Save mode or a Play mode, respectively.

Control of Variable-Speed Playback

A helical-scan VCR can play back pictures at any speed by changing the tape transport speed. However, since the rotating video heads cannot trace the recorded tracks accurately when the playback speed differs from the speed at which the recording was made, severe degradations in the playback picture will result. These occur as band-shaped noise (referred to as the *noise band*) and a vertical swaying of the picture.

Since only one noise band occurs in a given picture, it can be driven into the vertical blanking period if certain conditions are met. When the noise band is driven into the vertical blanking period, it disappears from the playback picture—a result accomplished by automatically adjusting the delay time of the tracking control signal in proportion to the tape speed.

When a noise band is driven into the vertical blanking period, however, the vertical sync is inevitably impaired, and vertical swaying occurs. In order to avoid this swaying, pseudo-vertical sync is added at an optimum phase (additional V sync).

To eliminate both noise band and vertical swaying, the following technologies may be applied:

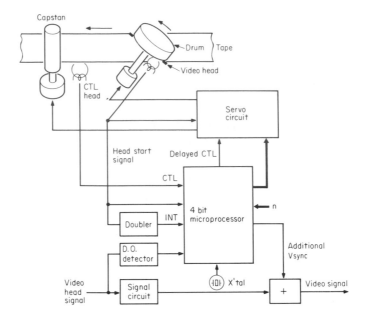

Fig. 3-18 Microprocessor control in variable speed playback (N. Azuma et al, Hitachi, Ltd, Consumer Electronics, © IEEE)

1. Tracking with offset to place the noise band in the vertical blanking period
2. Adding optimum pseudo-vertical sync signal to eliminate vertical swaying
3. Control of the tape stopping point to eliminate the noise band in the freeze-frame mode

A microprocessor-controlled system arranged as shown in Fig. 3-18 performs the following functions:

1. Delays the control signal (CTL) and feeds it into the servo circuit, thus accomplishing tracking with offset
2. Doubles the head start signal in accordance with the head drum rotation and delays it to provide the additional V sync needed to eliminate vertical swaying
3. Adjusts the drop-out position to the vertical head start position, thus eliminating the noise band from the still and freeze-frame modes

Since it is difficult for the microprocessor to process two independent pulses so as to delay both the CTL and the head start signal in a very short stretch of time, this particular design has provided for the interruption of the head start signal.

The TV Receiver in the Recorder

It should not be forgotten that most home TV recorders contain a complete TV receiver (minus the cathode ray tube and associated circuitry

Fig. 3-19 Panasonic's PV-1400 VHS recorder incorporates programming and search features as well as four video heads (*Courtesy*, Panasonic)

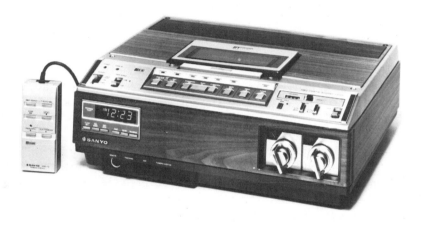

Fig. 3-20 Sanyo's Betamax Model 5050 incorporates a wide range of features as well as remote control of some functions (*Courtesy*, Sanyo Electric, Inc.)

needed for its operation). Although many recorder models incorporated detent-type electromechanical TV tuners, most of the more recent models utilize electronic push-button tuning of a varactor type that lends itself to more flexible programming of the recorder. With varactor tuning, various TV channels can be recorded at will over an extended period by programmed timing. In contrast, recorders with detent tuning can record only one channel (preselected by tuning knob rotation) at various intervals over an extended period.

The TV receiver section of the video recorder should be serviced in the same way as conventional TV receivers of similar design. Almost any one entering the video recorder service field will be adequately equipped to service its TV receivers because of prior work in that field. However, those just entering both fields should be aware of the advantages of test equipment applicable to both—for example, TV analyzers like the Sencore VA48, which was designed for both TV receiver and home video recorder servicing.

Remote Control of Home Video Recorders

As is true of TV receivers, remote control of home recorders is extremely desirable, particularly in deluxe models. Most remote controls are wired to the recorder and control tape speeds, rewind, and stop functions. If fast-scan and freeze-frame features are available, they will probably be remotely controlled as well. A most elaborate remote control for video recorders is the 15-function wireless unit available for use with Mitsubishi's model HS-300U.

Typical Popularly-Priced Models

Typical multi-featured home video recorders encountered in the service field offer most of the features mentioned above at prices that should result in their wide distribution.

A typical VHS recorder, Panasonic's PV-1400, shown in Fig. 3-19, provides 2-, 4-, and 6-hr record/playback times. Equipped with 14-pushbutton electronic tuning, the programmable tuner/timer can be set to turn the recorder on and off and change channels as many as eight times during a period of two weeks. The microprocessor control system of the tuner/timer can memorize the day of the program, the channel, when the program starts, and its duration. A feature called *Omni-Search* permits the user to see a recording on the TV screen as it runs in fast forward or rewind (in the 4-hr and 6-hr modes). Tape movement is at nine times the speed of conventional playback, forward, and reverse (cue and review). A stable picture is maintained during the fast-search operations. Additional features include audio dubbing capabilities, memory rewind, auto stop, and auto rewind at the end of the tape. Direct-drive capstan and cylinder motors are used to assure picture stability.

A Beta format unit, Sanyo's model 5050, is shown in Fig. 3-20. This recorder provides a wide range of features in an economical price range. Although there are several Beta models on the market with electronic tuning, Sanyo provides remote control in this model and forgoes pushbutton electronic tuning to keep the price within a certain range. In most other features, it offers a wide range within the Beta format, such as a maximum of 5-hr record/playback time and a scan speed of 15 times the normal rate for fast picture search. The tape-loading mechanism design of the Beta system makes possible higher search speeds than is possible with VHS machines, as well as freeze frame with single-frame advance. In other words, the action can be watched one frame at a time. Preprogramming is limited to one channel in a 24-hr period because of detent tuning.

A compact hand-held remote control unit allows the viewer to control Play, Rewind, and Stop functions, high-speed forward or reverse scan of pictures, and freeze frame with single-frame advance.

4

Video Signal Circuitry

In this and the next two chapters, the circuitry of the video section (Chap. 4), the servo system (Chap. 5), and the control system (Chap. 6), of a VHS recorder will be described in detail. In each case, the Hitachi model 5000A was chosen for discussion and is presented with the permission of Hitachi Sales Corp. of America.

Luminance Signal Recording System

The signal supplied to the video input is controlled by the AGC circuit in IC201 (Fig. 4-1) so that the output level is kept constant despite changes in the input level. After having its 3.58-MHz chrominance signal components removed and its bandwidth limited to 3 MHz by a low-pass filter LPF composed of Q204, CP201, and CP202, the video signal is fed to the pre-emphasis circuit. The pre-emphasis circuit emphasizes high frequencies prior to recording so that the signal-to-noise ratio will be improved during

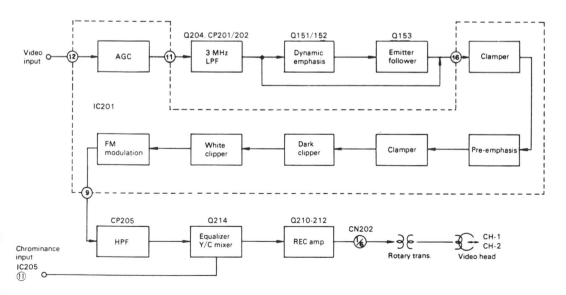

Fig. 4-1 Block diagram of luminance signal recording system

demodulation in the playback mode. Since frequency modulated signals are apt to be affected by noise at higher frequencies in general, high-frequency components are pre-emphasized to increase them relative to the noise. The pre-emphasized video signal is fed to the sync tip clamper, which fixes the dc potential of the sync tip of the synchronizing signal to prevent it from changing when there is a change in the picture contents of the video signal and therefore ensure that the FM modulator will hold the sync tip at 3.4 MHz.

The TV signal varies between the level of the sync tip and the level of peak white. Hence the limits of the frequency modulation deviation are set between a frequency representing sync-tip level at the lowest swing of the FM modulator and a frequency representing peak-white at the highest swing. If the FM modulator were not set at these limits, over-modulation would occur and the picture reproduction would be unsatisfactory. In VHS systems, the sync tip is set at 3.4 MHz, whereas in Betamax systems it is set at 3.5 MHz. In VHS systems, peak-white is set at 4.4 MHz and in Betamax systems at 4.8 MHz.

The setting of the sync-tip levels and peak-white levels of the FM deviation are simply fixed to restrict the FM swing to the limits necessary to insure a satisfactory recording and playback of the picture. However, as was shown in Fig. 2-15, the sidebands of the FM signal extend beyond the sync-tip and peak-white maximum deviation limits. In frequency modulation, sidebands are always generated beyond the deviation limits, and their amplitude depends upon the deviation ratio of the FM—that is, the ratio of the deviation frequency to the carrier frequency being modulated. In the demodulation process, it is necessary to recover the sidebands either above

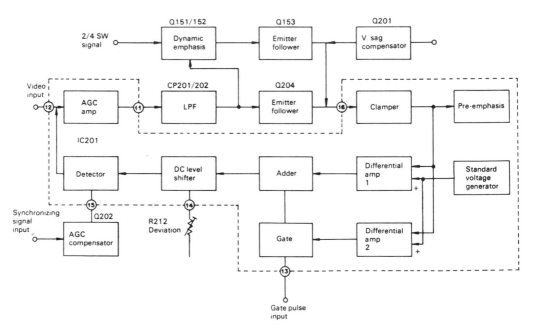

Fig. 4-2 Block diagram of AGC system

or below the carrier to ensure faithful reproduction. In the case of video recorders, the lower sidebands are recovered to avoid the extended frequency range that would result from the use of the upper sidebands.

The clamped signal is fed to the dark clipper and the white clipper, the functions of which are to clip over-shoots above specified levels. Without these circuits, sharp over-shoots occurring in the rise and decay sections of the video signal could so enlarge the instantaneous frequency deviations of the FM modulator that overmodulation would result. In playback, these effects could deteriorate the signal-to-noise ratio and degrade the picture.

The clipped signal is fed to the FM modulator, where it is modulated so that its sync tip is 3.4 MHz and the white peak is 4.4 MHz. After the attenuation of its chroma-band signal (629 kHz ± 500 kHz) by the high-pass filter HPF CP205, the FM-modulated signal is mixed with the low-pass chroma signal and amplified by the recording amplifier, consisting of Q210 to Q212. It is then fed to the video heads via the rotary transformers and finally recorded on the video tape.

The sections that which make up the Luminance Signal Recording System shown in Fig. 4-1 will now be described in greater detail.

AGC in Record Mode (Fig. 4-2)

The video input signal entering IC201 at pin 12 is fed to the sync-tip clamper through pin 16 and via the AGC amplifier through pin 11 to the low-pass filter LPF and the emitter follower Q204. In the sync-tip clamper, the synchronizing signal tip is fixed and transmitted to the pre-emphasis circuit. Also, it is fed to differential amplifiers (1) and (2). The gain of differential amplifier (2) is set to 7/3 of the gain of differential amplifier (1). The synchronizing signal obtained by gating during the back-porch period of the synchronizing signal in the gate circuit is added to the output of the differential amplifier (1) by the adder. The added signal passes through the dc level shifter and its peak value is detected. The gain of the AGC amplifier is controlled by this detector output. The AGC output is adjusted by controlling the amount of dc level shift using resistor R212 (the deviation control), which is connected to pin 14. Circuit Q201, which is connected to pin 16, acts to fully suppress the sag that is likely to occur in the clamp circuit. Circuit Q202, which is normally off, improves transient response when the synchronizing video signal is at a low level. It is seen connected to pin 15 in Fig. 4-2.

Pre-Emphasis

The video signal from the AGC output is fed through low-pass filter LPF (CP201, CP202), the characteristic of which is shown in Fig. 4-3. This filter acts to prevent beats between the FM modulated signal and the chroma signal by limiting the video signal bandwidth to 3 MHz. The video signal, after passing through the LPF, is subemphasized by Q151 and Q152 to compensate for the signal-to-noise ratio deterioration caused by the narrow track width in the LP mode. Signals of small amplitude are

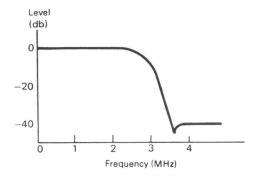

Fig. 4-3 Low-pass filter characteristics

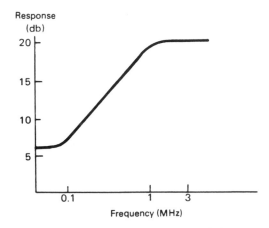

Fig. 4-4 Pre-emphasis characteristics

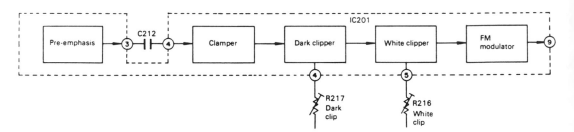

Fig. 4-5 Block diagram of clamper and clipper circuits

subemphasized much more. This action is referred to as *dynamic emphasis*. It is actually a subemphasis that precedes the main pre-emphasis circuitry in IC201.

Through emitter follower Q153, the signal is then fed back into IC201 via pin 16 to the main pre-emphasis circuits. The pre-emphasis characteristic shown in Fig. 4-4 is realized by the feedback circuit of the high-gain amplifier in IC201. Corresponding de-emphasis in the playback mode results in a greatly improved signal-to-noise ratio, particularly at the higher frequencies.

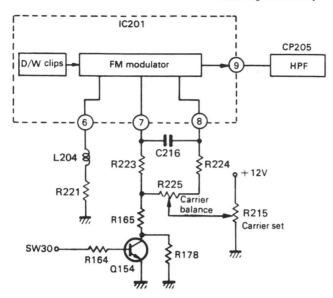

Fig. 4-6 FM modulation circuit

Clamper, White Clip and Dark Clip

The pre-emphasized video signal is fed to the clamp circuit through pin 3, C212 and pin 4 of IC201 to fix the sync tip to a certain potential and hold it at 3.4 MHz during frequency modulation. The clamped signal is given dark clip and white clip by circuitry in IC201. Adjustments of dark and white clip levels are made by R216 and R217, as shown in Fig. 4-5.

FM Modulation

After clipping, the video signal is fed to the FM modulator, which employs an ac emitter-coupled non-stable multivibrator. The oscillation frequency of this multivibrator, the circuit of which is shown in Fig. 4-6, is determined by the following formula:

$$f_o = \frac{\dfrac{Ve}{R221} + Is}{4K \times C216 \times Vcc}$$

where

K = constant determined by the ratio of resistances inside IC201 (value = 0.046)

Vcc = power voltage

Ve = potential at pin 6

Is = current flowing through R223 (value = 0.3 mA)

When modulated by the video signal, this unit is designed to operate as shown in Fig. 4-7 with the sync tip of the video signal at 3.4 MHz and with white peak at 4.4 MHz. The sync-tip frequency is adjusted by varying Is in the above formula, using R215. By adjusting the ratio of the current flowing through R223 and R224 with that through R225 (the carrier balance control), the second harmonic distortion of the FM signal is suppressed. A low-pass filter action is provided by L204, connected to pin 6, in order to suppress interference caused by higher video harmonics.

HPF

At this point, the frequency-modulated signal is fed to high-pass filter HPF CP205, which attenuates that portion of the lower sideband of the FM signal that overlaps the low-pass (or color-under) converted chroma signal. The HPF characteristics above the chrominance signal band are shown in Fig. 4-8.

Recording Amplifier, Squelch and Y/C Mixer

The FM signal, having passed through the HPF just mentioned, is fed to the recording amplifier (Q210 to Q212), after having been mixed with the

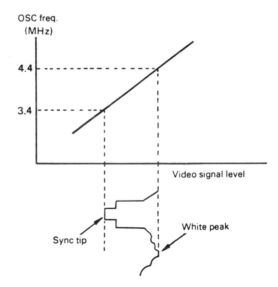

Fig. 4-7 FM modulation from sync tip to white peak

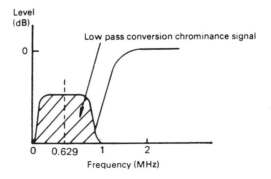

Fig. 4-8 High-pass filter charactristics

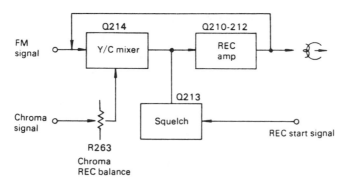

Fig. 4-9 Block diagram of Y/C mixer and squelch

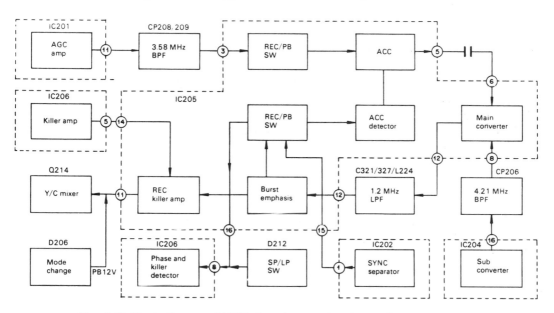

Fig. 4-10 Block diagram of VHS chrominance signal recording system

low-pass converted chroma signal in the Y/C mixer Q214 (see Fig. 4-9). The recording amplifier operates so that the recording current is optimized for all signal frequencies. In this recorder, it is a pure, complementary, single-ended push-pull circuit. Push-pull operation works well since little load current is required. Crossover distortion and second harmonic distortion are extremely low. Q213 provides squelch by preventing the signal from being fed to the recording amplifier for approximately 3 sec after loading. This circuit prevents the recorded signal from being erased when the tape starts running near the head drum before the loading operation is completed.

Chrominance Signal Recording System

An overall diagram of the chrominance-signal recording system is shown in Fig. 4-10. The 3.58-MHz chroma signal, which has been fed to the

3.58-MH bandpass filter BPF from the video signal of the AGC output (from IC201), is fed to the ACC (automatic color control) circuit via the receive/playback switch (REC/PB input signal changeover) in IC205. The chroma signal, which is controlled by the ACC to a certain level, is fed to the main converter. In the main converter, the 3.58-MHz chroma signal is mixed with the 4.21-MHz signal from the sub-converter in IC204, and frequency-conversion is performed. The differential-frequency component (629 kHz ± 500 kHz) is taken out via the 1.2-MHz low-pass filter and then fed to the REC killer amplifier via the burst-emphasis circuit. Only the burst signal is emphasized by 6 dB in the burst-emphasis circuit; it is fed to the Y/C mixer (Q214). The signal output from REC/PB switch SW is applied to the phase detector of IC206.

The sections that make up the Chrominance Signal Recording System shown in Fig. 4-10 will now be described in greater detail.

The Band-Pass Filter (3.58-MHz BPF)

The video signal fed from the AGC output of IC201 is taken out as the 3.58-MHz chroma signal with a bandwidth of ± 500 kHz to the 3.58-MHz BPF. Band-pass characteristics of this BPF are shown in Fig. 4-11.

ACC (Automatic Color Control)

The ACC circuit in IC205 is used for both recording and playback. Since power is not supplied to IC203 (the main IC involved in playback operation) during recording, the playback signal is not fed to pin 4, and only the 3.58-MHz chroma signal is fed to the ACC through pin 3 via the input-signal switching circuit in the record mode. The 3.58-MHz chroma signal at ACC output pin 5 is then fed to pin 6. When the signal is fed to the recording-system burst-gate amplifier before being frequency-converted and burst-emphasized, only the burst signal is selected as the ACC detection input. This detector input is comparison-controlled by the detection level and maintains the burst-signal level constant at pin 5.

The Main Converter

The main converter frequency-converts the chroma signal (3.58 MHz ± 500 kHz) from the ACC to the low-pass conversion chroma signal (629 kHz ± 500 kHz) for color-under operation. The chroma signal input from pin 6 is mixed with the 4.21-MHz carrier signal, which is fed through the IC204 sub-converter via pin 8 and delivered to pin 9 as the sum and difference frequency components. The sum component is 7.78 MHz ± 500 kHz, and the difference component is 629 kHz ± 500 kHz. These components are fed to the 1.2-MHz low-pass filter, which passes the lower frequency component of 629 kHz ± 500 kHz. The characteristics of the 1.2-MHz BPF are shown in Fig. 4-12.

Killer Amplifier and Burst Emphasis

The low-pass converted chroma signal (629 kHz ± 500 kHz) taken out through the converter output pin 9 via the 1.2-MHz LPF is fed to the burst-

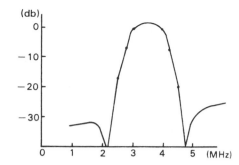

Fig. 4-11 3.58-MHz band-pass filter characteristics

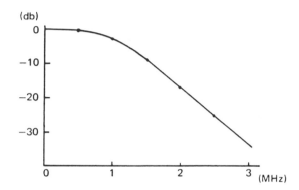

Fig. 4-12 1.2-MHz low-pass filter characteristics

emphasis and color killer amplifier through pin 12. Only the burst period is fed out through pin 11 after being emphasized by 6 dB. The color killer amplifier, at that time, suppresses any leakage from the 3.58-MHz component contained in luminance (black/white) signal output through pin 11 during the black/white signal period.

Mode Selector

IC205 is used in both recording and playback. Mode selection is performed by applying 12 V during playback and 0 V during recording to the mode selector terminal (D206 anode). The recording chroma-signal output pin 11 is at high potential during playback, and the recording killer amplifier is turned off. This action suppresses the recording output. At the same time, the playback control signal is generated in IC205, and the respective switching circuits are operated in the playback mode. LP/SP (long play/standard play) changeover is such that burst emphasis is performed during SP and not during LP. During SP, 0 V is applied to D212 and R332, and 12 V is applied during LP.

Luminance/Chrominance Signal Playback System

The playback signals from CH-1 (channel 1) and CH-2 (channel 2) video heads are applied via the rotary transformer to separate pre-amps

(IC203 pins 11 and 9) for CH-1 and CH-2 and amplified. These outputs are fed to the channel switch, where the circuit switches the signals from the two heads by a 30-Hz pulse fed from the servo circuit to form a series signal. Part of this output is taken from pin 6 of IC203 and fed to chroma playback amplifier Q220 via buffer amplifier Q219. After being amplified further, it is fed to the low-pass filter (consisting of C337 to C339 and L228/229), and the low-pass converted chroma signal component is taken out to be fed to the chroma playback system. The other output is fed to the AGC. This circuit absorbs the inter-channel level deviation of CH-1 and CH-2 video playback signals and thus keeps the output level constant. Defects in the FM signal—dropouts, which occur as a result of scratches, etc., on the magnetic tape—are detected by the dropout (DO) detector and corrected by the circuit consisting of the 1-H delay line (DL201), the DO switch, and adder (in IC203). This output is fed through pin 20, where a high-pass filter (HPF) and the phase equalizer (Q218 and Q222) are connected. There, the FM band component is selected to improve the frequency response. After improving the phase characteristics in the FM phase equalizer, the signal is fed to the input of the inversion-preventer (in IC203) via the buffer amplifier (Q223). This circuit prevents the *inversion phenomenon*, and the next limiter eliminates the AM component of the FM signal. Buffer 217 feeds the signal to the frequency-demodulation system.

The magnetic record/playback characteristics of FM signals are such that the higher the frequency, the higher the noise. To compensate for this characteristic, the signal is pre-emphasized at the higher frequencies prior to recording in order to place them at a level above the noise; the greater the pre-emphasis, the greater the improvement in signal-to-noise ratio. However, the presence of pre-emphasis will cause a sharp video signal rise or fall to produce a sharp overshoot or undershoot of voltage, resulting in overmodulation of the FM signal. Under such conditions, the lower sideband amplitude of the FM signal increases, and AM components result when demodulated to produce a negative-conversion picture. This effect is described as "inversion phenomenon" or "black and white reverse phenomenon." The result is also an extreme signal/noise ratio deterioration in this portion of the picture. Hence, the circuitry provided to correct for this situation is known as an "inversion-preventer" or an "S/N improver."

The sections that make up the Luminance/Chrominance Signal Playback System shown in Fig. 4-13 will now be described in greater detail.

Pre-Amplifier

Since the signal level obtained from the tape via the video heads is extremely small, it is amplified to approximately 0.2 V peak-to-peak by pre-amplifiers CH-1 and CH-2. At the same time, the high-pass response is improved. The signal-to-noise ratio must be improved by this first-stage amplification, and the matching of the video heads must be taken into account. The pre-amplifier has in its first stage a cascade of low-noise transistors to improve the S/N ratio. At the same time, feedback control is employed for damping. Since the playback frequency response falls off

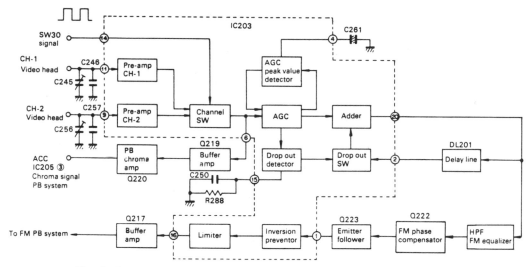

Fig. 4-13 Block diagram of VHS luminance/chrominance playback system

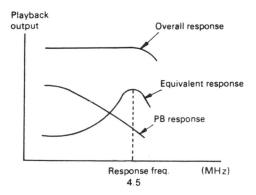

Fig. 4-14 Playback frequency response

toward the high frequencies, that response is compensated ahead of the pre-amplifier with the resonance of the L component of the heads and the C components C245 and C256 to obtain an overall flat frequency response (see Fig. 4-14). L209 is connected to IC203 pin 5 to perform additional peaking. The pre-amplifiers for Channel-1 and Channel-2 are identical; their outputs are supplied to the channel switch described below.

Head Switching

In the helical scanning system using two video heads, signals from the heads are switched to make one series signal. In general, the tape-winding angle is made a little more than 180 deg so that the Channel-1 and Channel-2 signals overlap. Simply to mix them creates problems. A procedure for mixing them properly is shown in Fig. 4-15. The playback signals from CH-1 and CH-2 enter alternatively every 180 deg with 30-Hz synchronization. Noise is contained in the nonsignal section, and overlap-

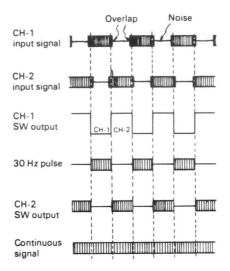

Fig. 4-15 Timing chart of video head switching

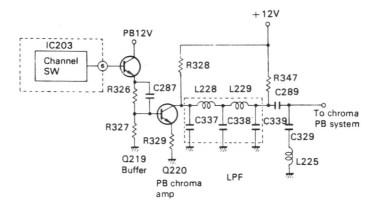

Fig. 4-16 Schematic diagram of buffer and playback chroma amplifier

ping is provided to prevent picture loss. Thus, CH-1 and CH-2 cannot be mixed simply. Their signals must be mixed with the 30-Hz pulses from the servo circuit (as shown in Fig. 4-15) to take out clean signals without overlap or noise. These signals must be mixed to obtain a series playback signal of the proper structure. Part of this signal is fed to the AGC and the other to the playback chroma amplifier via the buffer through pin 6 of IC203. It must be pointed out that the polarity of the 30-Hz pulse must be specified as shown in Fig. 4-15. If the polarity is inverted, switching cannot be accomplished and the signal cannot be obtained.

Buffer Amplifier and Playback Chroma Amplifier

The signal from IC203 pin 6 is amplified by playback chroma amplifier Q220 via buffer Q219 and fed to the low-pass filter consisting of C337/338 and L228/229 (see Fig. 4-16). In this circuit, the low-pass converted chroma signal is taken out of the playback signal. The chroma signal thus obtained is fed to the chroma signal playback system. The characteristic impedance of the LPF is 2.2 k ohms, and the output matches the input resistance of the chroma signal playback IC. The circuit composed of L225 and C329 is a trap to prevent audio bias from interfering in audio dubbing recording.

AGC in Playback Mode

When there is inter-channel deviation between the playback output of the video heads, undesirable phenomena such as flicker, etc., occur, and consequently a balance adjustment of the playback FM channel is required. When the playback output becomes uneven because of unevenness in the characteristics of the video heads, it causes problems in later FM circuits, making FM playback level adjustment necessary. In this design, AGC is employed in the FM playback system automatically to compensate for both items.

Part of the signal of the channel switch output is fed to the AGC. The AGC output is fed to the AGC detector, and its peak value is detected. The ac component is removed by C261; it is converted to dc and applied as the AGC control signal. Capacitor C261; which is connected to pin 4, sets a time constant and responds sufficiently quickly to level changes every one-sixtieth of a second. However, it does not respond to short period changes such as dropout—that is, to detect dropout precisely. Figure 4-17 shows the input and output waveforms of the AGC. A signal with approximately 1 V peak-to-peak is obtained at the AGC output (pin 20 of IC207).

Dropout (DO) Detector, Dropout (DO) Switcher, and Adder

Part of the AGC output is fed to the DO detector, where it is given double sideband rectification by the limiter. Then its ac component is removed by C250, which is connected to pin 15, and converted to a dc voltage corresponding to the AGC output amplitude. When dropout occurs, the voltage at pin 15 falls, and the level of the voltage is detected. Also, the dc switch closes. At that time, the signal, which is always input to pin 2 via the 1-H delay line, is added to the original signal. The addition of the delayed 1-H signal compensates for the line that has the dropout. *Hysteresis* is provided at the On and Off levels of the DO switch in this unit in order to prevent erroneous detection resulting from noise near the detection level. L211 and L212 are used for matching delay-line filter DL201.

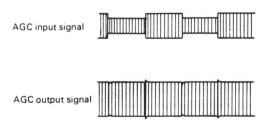

AGC input signal

AGC output signal

Fig. 4-17 AGC input and output signals

The term "hysteresis" as used here does not refer to magnetic hysteresis but is used in a broader sense. It refers to the functioning of the Drop-Out Compensator (DOC), where a lost line of the picture is replaced by the line that occurred before the lost line occurred. In a sense, this is a hysteresis effect in that a signal from the past is retained to substitute for the current one, which has been lost.

HPF, FM Phase Equalizer (Q218, Q222), and Buffer Amplifier (Q223)

The AGC output signal is fed to the high-pass filter in the input circuit of Q218, and the FM band component is removed. Frequency response is improved in the next FM equalizer. This signal, with its phase characteristics improved in the next FM phase compensator (Q222 following Q218), is again fed to IC203. There at pin 1 it enters the FM inversion preventor via buffer amplifier Q223. The FM phase-equalizer circuit is shown in Fig. 4-18.

Inversion Preventor

The inversion phenomenon has previously been discussed. In this design, the recording wavelength is not used for high-density recording, but more pre-emphasis is applied. Therefore, when the FM carrier is less than the low-pass sideband, the inversion phenomenon is likely to occur. The inversion-prevention circuit is shown in Fig. 4-19.

At normal input level where inversion does not occur, the diode becomes conductive and limiting is applied. When the level of the FM carrier is lowered and the inversion phenomenon is likely to occur, the diode is opened. In addition, the circuit composed of R293 and L213 is provided with high-pass filter characteristics, the low-pass sideband is suppressed, and the normal composite FM wave is obtained. By feeding this signal to the next limiter, inversion is prevented. The signal, which has

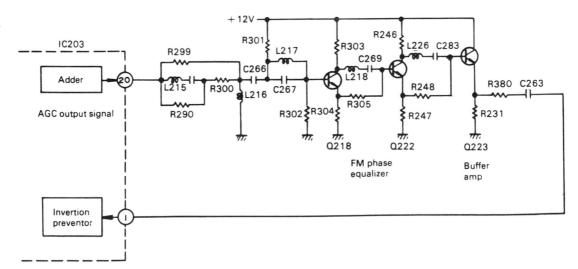

Fig. 4-18 FM phase equalizer circuits

had its AM component removed by the limiter, is taken out from pin 16 and fed to the frequency-demodulation system via buffer amplifier Q217.

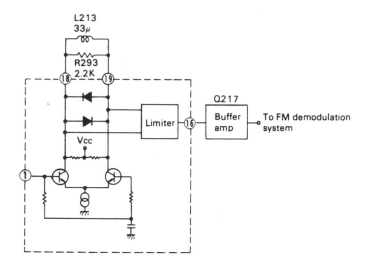

Fig. 4-19 Inversion prevention circuit

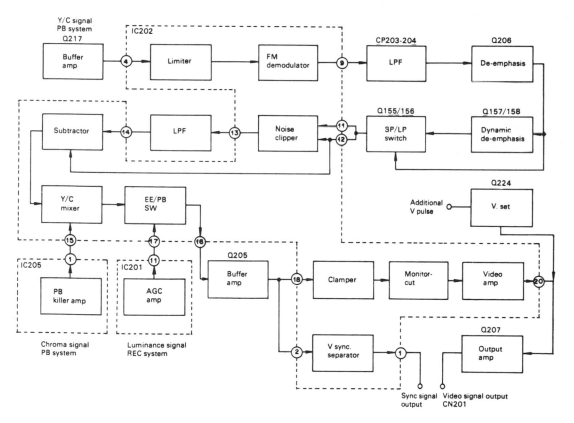

Fig. 4-20 Block diagram of VHS luminance signal playback system

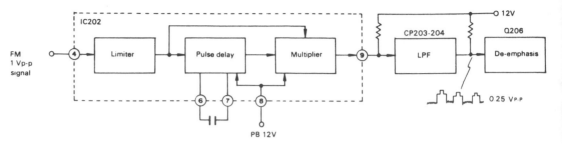

Fig. 4-21 Block diagram of limiting and demodulation functions

Luminance Signal Playback System

To continue with the processing of the luminance and chrominance signals in the playback mode, separate diagrams will be shown for these signals in detail. The luminance signal playback system leading through a limiter into the FM demodulator is shown in Fig. 4-20. Following the FM demodulator are the noise canceller, Y/C mixer, video-signal selector circuit, monitor-cut and synchronization separator circuit (all of which are integrated in IC202). A low-pass filter, de-emphasis circuit, and a 75-ohm output amplifier are installed externally to IC202. (The chrominance signal playback system will appear in Fig. 4-26.)

Limiter, FM Demodulator, and Low-Pass Filter (CP203 and CP204)

An FM signal of approximately 1 V peak-to-peak is fed to the limiter; being limited later by a differential amplifier with a gain of about 46 dB, it is returned to a video signal by the FM demodulator (see Fig. 4-21). The latter is composed of a pulse-delay circuit and multiplier so as to obtain a positive synchronizing signal of approximately 250 mV peak-to-peak at the LPF output. The usual limiter-balance controls and carrier-balance controls have been removed by enclosing the circuit in an IC and by improving the off-band characteristics of the low-pass filter. This LPF is the same type as that used in the recording circuit.

Figure 4-22 shows the characteristics of the FM demodulator. The demodulation characteristics are determined by the delay time τ of the pulse-delay circuit and the current flowing through pin 8 of IC202. The demodulation has the following design parameters:

$$\tau \propto \frac{C}{I} \qquad \tau = 60 \ \mu s$$

$$f_o = \frac{1}{2\tau \times 10^{-3}} \qquad f_o = 8 \ \text{MHz}$$

Current flowing to pin 8 is supplied from outside the IC to keep any unevenness in demodulation characteristics to a minimum, to detect the

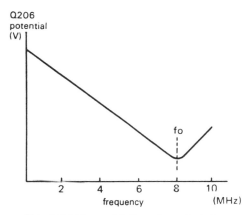

Fig. 4-22 Demodulation characteristics

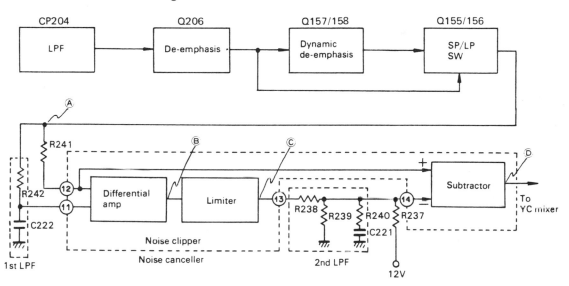

Fig. 4-23 Block diagram of noise canceller

presence or absence of this current, and to facilitate the changeover of IC operation in the playback and other modes.

Output Level Adjustment, De-emphasis, and Noise Canceller

Q206 is installed to match impedances between low-pass filter CP204 and the de-emphasis circuit and to adjust the playback video level (see Fig. 4-23). De-emphasis output is 1 volt peak-to-peak signal with negative synchronization. The noise canceller is composed of the differential amplifier (which amplifies the difference between the video signal and the video signal passing the LPF composed of R242 and C222); the limiter; the LPF composed of R238, R239, R240, and C221; and the subtractor. In Fig. 4-24, the high-pass component and noise in the video signal shown in Fig. 4-24(B)

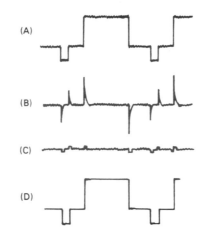

(A)

(B)

(C)

(D)

Fig. 4-24 Waveforms of noise canceller

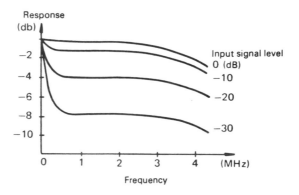

Fig. 4-25 Characteristics of the noise canceller

are removed by the filter (designated *1st LPF*) and the differential amplifier and clipped by the limiter to produce the waveform shown in Fig. 4-24(C). This waveform passes through the filter (designated *2nd LPF*) and is fed to the subtractor to be subtracted from the original video signal producing the waveform of Fig. 4-24(D).

Thus, it is seen that most of the high-pass component—which is mostly noise, of small amplitude, and not clipped by the limiter—is almost completely cancelled. However, the high component (the outline of the video signal) is clipped by the limiter, and almost nothing is subtracted. Thus, only noise is removed without any loss of picture quality. The role of the 2nd LPF in the noise canceller is to make any remaining noise in the waveform inconspicuous. The frequency response of the noise canceller is shown in Fig. 4-25. It is seen that its frequency response is dependent on the input level.

Y (Luminance)/C (Chrominance) Mixer and EE/Playback Selector

Y/C mixing is performed by adding the playback chroma signal to IC202 pin 15 and by capacitor-dividing the input circuit of the chroma signal

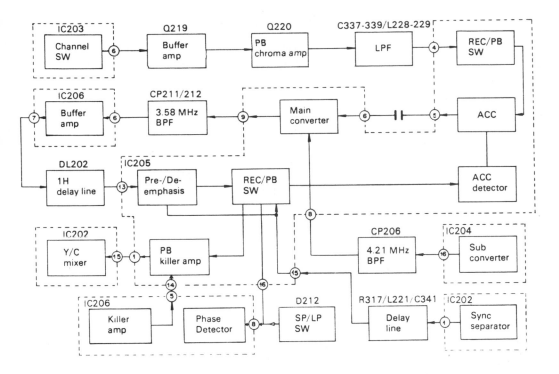

Fig. 4-26 Block diagram of VHS chrominance signal playback system

to improve the frequency response of the Y signal. It is designed so that when the EE video signal is applied to pin 17 and the dc potential at pin 8 is high, the output at pin 16 is changed over to the playback video output; when the dc potential is low, the output is changed to the EE video. In addition, it is designed so that the dc potential falls by approximately 0.5 V when the output of pin 16 is EE when compared with that during playback. This provision facilitates clamping at later stages when changing over from playback to EE.

Monitor Cut and Synchronization Separator

Monitor cut is designed to operate during playback and whenever high potential is applied to pin 19; it functions so that the picture does not appear for a few seconds after the playback button is pressed—the amount of time needed to present a stable picture. The dc potential during monitor cut at pin 20 is set to be the same as the gray color level of the video signal.

The sync separator operates in both the recording and playback modes. It feeds to the Y, the chrominance, and the servo systems during recording and to the chrominance system during playback.

Chrominance Signal Playback System

Figure 4-26 starts with the signal from the channel switch in IC203 and includes the buffer amplifier Q219 and the playback chroma amplifier Q220 shown in Fig. 4-13. After passing through these two amplifiers, the playback chroma signal is fed to the record/playback switch in IC205 through pin 4 as a 629 Hz ± 500 kHz low-pass converted signal, but only during playback. The recording 3.58-MHz chroma signal is suppressed in the switch, and only the 629-kHz chroma signal is fed to the ACC (Automatic Color Control). The signal, which has been controlled to a certain burst level by the ACC, is fed to the main converter and then frequency-converted with the 4.21-MHz carrier signal subconverter in IC204. It is further fed to the 3.58-MHz band-pass filter, from which the difference frequency signal is taken out.

This signal is fed to the 1-H delay-line filter (DL202) and the phase detector of IC206 via the buffer amplifier in IC206. The 1-H delay-line filter removes the crosstalk component of the chroma signal and improves the signal-to-noise ratio. Then the signal is fed to the pre/de-emphasis and the following record/playback switch.

In the color burst de-emphasis circuit, the burst signal emphasized by 6 dB during recording is attenuated by 6 dB during standard play, and the processed chroma signal is fed to the Y/C mixer in IC202. Since burst emphasis is not performed in the long play mode during recording, switching is required to prevent the de-emphasis circuit from operating in that mode. From the record/playback switch, following the pre/de-emphasis, only the burst signal is taken out and fed to the ACC detector.

Buffer Amplifier Q219 and Playback Chroma Amplifier Q220

Buffer amplifier Q219 operates only during playback. The playback chroma FM signal from the channel switch output of IC203 is taken out as a 629-kHz low-pass converted chroma signal during playback only and is fed to the playback/record switch in IC205.

Record/Playback Switch

The 3.58-MHz chroma signal is fed to pin 3 (shown in Fig. 4-13) regardless of whether the record or playback mode is being used. For this purpose, the 3.58-MHz chroma signal is fed to the ACC through pin 3 during recording and the 629-kHz chroma signal is fed through pin 4 during playback. This action is controlled by the Record/Playback switch (upper right-hand corner of Fig. 4-26).

ACC (Automatic Color Control)

The ACC output of the 629-kHz chroma signal is frequency-converted to 3.58 MHz via the main converter circuit during playback and is fed to the

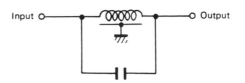

Fig. 4-27 1H delay line equivalent circuit

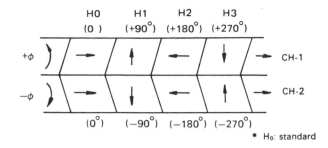

Fig. 4-28 VHS recording chroma vector relationships

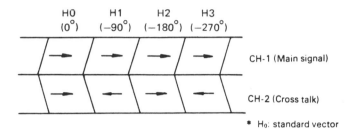

Fig. 4-29 VHS playback chroma vector relationships

playback-system burst-gate amplifier with the burst still emphasized. Only the burst signal is taken out at this point and fed to the ACC detector. Accordingly, the burst-signal input to the ACC detector is the 3.58-MHz signal in both the recording and playback modes, and the input burst level of the burst de-emphasis is held constant during playback.

1-H Delay-Line Filter

The 1-H delay-line filter is designed to allow only the 3.58-MHz chroma signal component to pass. Its equivalent circuit is composed of a 1-H delay element and a straight by-pass line, as shown in Fig. 4-27. In the VHS color-recording system, the channel-1 track is recorded with the chroma phase advanced by 90 deg every 1 H, and the channel-2 track is recorded with the chroma phase delayed by 90 deg every 1 H. This recording pattern is shown in Fig. 4-28. The result of the channel-1 video head tracing the recording pattern shown in this figure while overlapping a little with channel 1 is shown in Fig. 4-29. As shown in this figure, the vector indication of the channel-1 main signal is always kept at H_o (the reference of the delay line) by means of AFC operation, and the vector indication of the channel-2 crosstalk component varies by 180 deg every 1 H.

Accordingly, by feeding this signal to the 1-H delay line, the main signal component becomes a composite vector (approximately double in value), and the crosstalk component is cancelled. In other words, the crosstalk from the adjacent track is removed by this 1-H delay line. In so doing, the S/N ratio of the chroma signal is improved by 3 dB by this 1-H delay line.

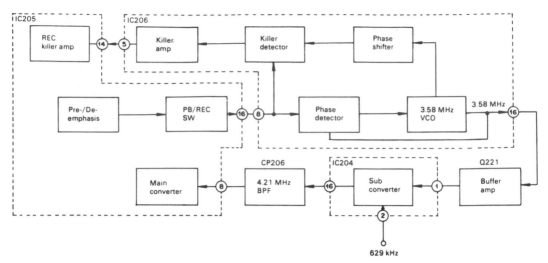

Fig. 4-30 Block diagram of VHS chrominance signal recording APC system

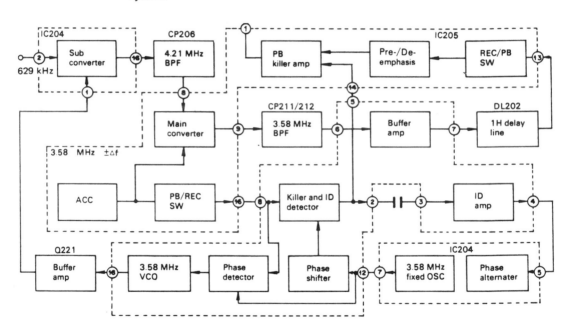

Fig. 4-31 Block diagram of VHS chrominance signal playback APC system

Burst De-emphasis

Since only the burst portion of the signal is emphasized by 6 dB by pre-emphasis in the SP (Standard Play) record mode, it is required to attenuate only the burst signal by 6 dB during playback to return to the original level ratio. The 3.58-MHz chroma signal input to pin 13 from DL202 attenuates the amplifier gain by 6 dB during the period when the gate pulse is not applied (burst period)—attenuates it, that is, with respect to the period when the gate pulse is not applied (chroma period)—by means of the burst de-emphasis circuit, which is controlled by the burst gate pulse from pin 15. Burst emphasis and de-emphasis do not operate in the LP (long play) mode.

Color Killer Amplifier

The color killer function is provided in the de-emphasis circuit in the playback system in the same way as in the recording system. A dc voltage is supplied to pin 14 from IC206 during the chroma signal period. There it becomes 0 V during the white/black signal period to control the killer amplifier; the output from pin 1 is suppressed during this black/white period.

APC (Automatic Phase Control) System

Figure 4-30 is a block diagram of the Chrominance Signal Automatic Phase Control system in the record mode. Those portions of IC204, IC205, and IC206 used to accomplish the automatic phase control are shown as well as the main external blocks.

APC During Recording

During recording, the APC system serves two functions: (1) It produces the 3.58-MHz continuous wave and feeds it to the sub-converter, and (2) it generates the color-killer voltage. It is necessary for the APC system to operate if the color-killer function is to be performed during recording. In Fig. 4-30, the closed circuit composed of the phase detector and the 3.58-MHz VCO (Voltage-Controlled Oscillator) is the same as that in a TV receiver. The 3.58-MHz signal, which is produced by the phase detector and the 3.58-MHz VCO, synchronizes with the burst signal and is fed to the sub-converter. Simultaneously, the killer detector phase-compares the burst signal and the 3.58-MHz VCO output that is synchronized to the burst signal and generates a high voltage when the burst signal is normal phase and a low voltage for the other phases. The killer-voltage output is fed to the recording killer amplifier; the recording chroma signal can pass through the killer amplifier only during high-voltage periods when it is superimposed on the frequency-modulated Y signal and recorded on the tape.

APC During Playback

During playback (see Fig. 4-31), the APC system has three functions, as follows:

1. Phase control of chroma signal: It removes the time base error of the chroma signal, which could not be removed entirely by the color AFC.
2. Color-killer control by phase detection: It detects the playback burst signal and generates the color-killer voltage.
3. AFC control by ID (identification) pulse generation: It detects phase-rotation timing errors, which occur at the field changeover point, and generates ID pulses.

Phase control of chroma signal: The frequency of the playback chroma signal varies as it is affected by jitter, etc.

The time base error due to "jitter", $\pm \Delta f$, is caused by small changes in video head velocity. Such changes, which can be particularly disturbing in the case of color TV signals, must be removed in order to display satisfactory color pictures. Assuming that a playback chroma signal with a frequency of 629 kHz $\pm \Delta f$ is input to the main converter and that (considering the first stage) the APC system does not work, the 3.58-MHz VCO will generate a 3.58-MHz signal that is increased to 4.21 MHz by means of the subconverter and the 4.21 band-pass filter; this enlarged signal is then fed to the main converter. Accordingly, difference components of 629 kHz $\pm \Delta f$ and 4.21 MHz are taken out by the main converter and the 3.58-MHz band-pass filter. In other words, 3.58 MHz $\pm \Delta f$ is taken out and fed to the phase detector, killer ID detector via the buffer amplifier. At the same time, 3.58 MHz has been fed to the phase detector from the fixed 3.58-MHz oscillator. After being phase-compared with the burst component of the playback chroma signal, it feeds to the VCO the corresponding difference frequency.

Accordingly, 3.58-MHz $\pm \Delta f$ is generated in the VCO output and fed via the sub-converter 4.41-MHz band-pass filter to the main converter, where it is frequency-converted. The playback chroma signal frequency, which has passed the 3.58-MHz bandpass filter, is then calculated as follows: (4.21 MHz $\pm \Delta f$) $-$ (629 kHz $\pm \Delta f$) = 3.58 MHz. Thus it is seen that the variation component Δf in the playback chroma signal is cancelled by the function of the APC system, and a precise and constant 3.58-MHz ($\pm$ 500 KHz) chroma signal is obtained.

Color killer control: The burst component of the playback chroma signal and the 3.58-MHz signal from the fixed 3.58-MHz oscillator with 90-deg phase shift are phase-compared. When the phase is normal, a high voltage is generated and fed to the killer amplifier (IC205 pin 14).

AFC control: The killer detector is used as an ID detector, and when the phases of the two input signals differ by 180 deg, a positive polarity pulse is generated at pin 4 of IC206. By feeding this pulse to AFC IC pin 5, the playback chroma signal phase is returned to normal.

3.58-MHz VCO

In the phase detector, the difference voltage corresponding to the phase difference between the burst signal and the sub-carrier (3.58-MHz VCO during recording and fixed 3.58-MHz signal during playback) is fed to the VCO. The VCO delivers the signal, which has the proper phase to correct the phase deviation by means of the detection output difference voltage, and feeds it to the sub-converter. A schematic diagram of the VCO and a vector diagram showing the relation of phases between waveforms at all sections are shown in Figs. 4-32 and 4-33, respectively.

C306 and C307 are chosen so that V_3 is delayed by 135 deg with respect to V_1. In Fig. 4-33, V_1 shows when the detection output difference voltage is 0 V; when the difference voltage becomes positive or negative, the addition ratio of V_4 and V_5 varies. The V_1 phase varies depending on the

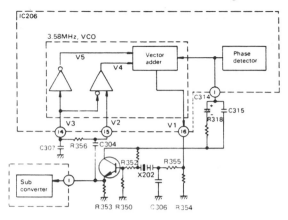

Fig. 4-32 3.58-MHz VCO and associated circuits

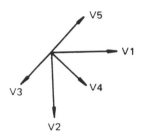

Fig. 4-33 Phase relationships in VCO circuitry

detection output difference voltage in the range of 0 ± 45 deg. This phase change changes the VCO oscillation mode and is fed to the sub-converter as a frequency change.

90-Deg Phase Shifter

It is necessary to keep the sub-carrier, which is fed to the phase detector and the color-killer detector, at a 90-deg phase difference (see Fig. 4-34). For that purpose, the phase difference of 90 deg is obtained in the following manner: The 3.58-MHz VCO output during recording and the fixed 3.58-MHz signal during playback are fed to IC206 pin 10 via the switch inside the IC. This signal is phase-shifted by approximately 45 deg by means of the externally attached R364 and C313 and fed to pin 9. The pin 9 input signal is fed to the killer detector from both ends of C313, and the difference signal between pin 10 input and pin 9 input (that is, the signals from both ends of R364) is fed to the phase detector. Accordingly, these two signals are always kept with a phase difference of 90 deg.

The killer-detector-input sub-carrier inverts during recording and playback by virtue of the inverter shown in the block to its left, and the phase of the phase-detector-input sub-carrier is changed over to 90-deg advanced phase and 90-deg delayed phase because the APC loop differs in recording and playback and synchronizes with the burst signal during

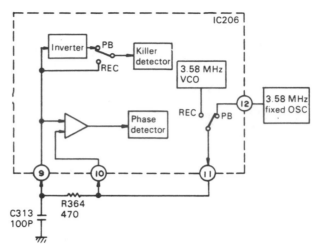

Fig. 4-34 90-deg phase shifter

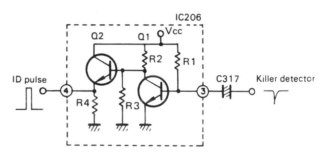

Fig. 4-35 ID pulse generator

recording. The system synchronizes with the fixed 3.58-MHz signal during playback.

ID (Identification) Pulse Detector

In this design, the killer detector is used simultaneously as the ID detector. The ID pulse generator provided generates ID pulses as shown in Fig. 4-35. Transistor Q1 in IC206 is always saturated by base current that is supplied via R1, and pin 4 is at 0 V. Assuming that the negative pulse shown in the diagram is generated in the killer detection output, the current flowing through R1 flows out via C317. When Q1 is released from saturation, the positive-polarity pulse that appears in the output is the ID pulse. In other words, when the signal shown in the diagram appears at the killer detection output because of the phase of the burst signal and sub-carrier, the above-mentioned operation is performed by this circuit and the ID pulse is delivered.

AFC (Automatic Frequency Control) System

The AFC system shown in Fig. 4-36 operates in the same way during both recording and playback. However, the horizontal synchronization of

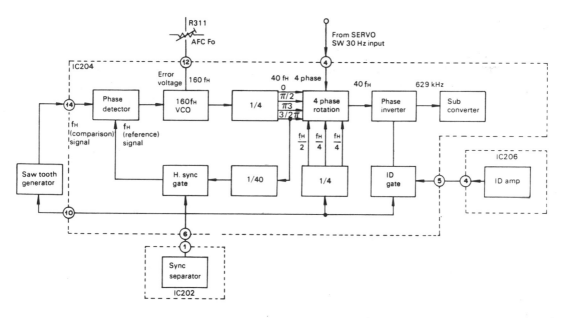

Fig. 4-36 Block diagram of automatic frequency control system

the input video signal is the AFC standard during recording, and the horizontal synchronization of the playback video signal is the AFC standard during playback. This system performs the following functions:

1. It generates a 40-f_h (horizontal sync frequency) low-pass carrier wave
2. It makes this 40-f_h wave rotate every 1 H
3. It compensates the phase control of the chroma signal by receiving ID pulses during playback.

Horizontal synchronization, which is to be the standard, is fed to the horizontal synchronizing gate circuit (pin 6 of IC204) from the synchronization separator (pin 1 of IC202) of the Y circuit, and there it is gated by the signal from the $\frac{1}{40}$ frequency divider. Unnecessary components (noise, etc.) are removed here. The resulting horizontal sync signal, from which unnecessary components have been removed, is fed to the phase detector. After this phase detector compares the phases of the reference horizontal synchronization and the signal from the $\frac{1}{40}$ frequency divider, it feeds the output to the 160-f_h VCO. The oscillation frequency of the VCO is controlled by the output of the phase detector. As can be seen from the block diagram, this VCO output is $\frac{1}{160}$ divided via the $\frac{1}{4}$ frequency divider and the $\frac{1}{40}$ frequency divider and is fed back to the phase detector. Therefore, to make the frequency of the $\frac{1}{40}$ division output (one input to the phase detector) the same as that of the reference horizontal sync (namely, f_h, another input) in this feedback circuit, the VCO oscillates at 160 times the frequency (namely, 160 f_h). This 160-f_h signal—VCO output—is fed to the $\frac{1}{4}$ frequency divider and becomes 40 f_h. This output is fed to the $\frac{1}{40}$ division and the four-phase rotation circuit so that 40 f_h is divided by 40 and f_h is generated.

This f_h output is divided into four parts. One part is fed to the $\frac{1}{40}$ frequency divider and is obtained as $f_h/2$ and $f_h/4$. Another part is fed to the horizontal sync gate as described previously. A third part becomes the comparison f_h of the phase detector via the saw-tooth wave-generation filter, and the fourth part is fed to the ID gate. There it gates the ID pulse and removes unnecessary pulses generated at positions other than the burst signal position, where the ID pulse should originally be generated.

The four-phase rotation circuit is the logic circuit, which rotates the phase of the $\frac{1}{4}$ division output—40 f_h—by 90 deg every 1 H by means of another $\frac{1}{4}$ division output—$f_h/2$, $f_h/4$—and SW30 Hz from the servo-printed circuit board.

The phase inverter inverts the phase of the four-phase rotation circuit output by the ID pulse, inverts the output color phase, and helps APC control of the output color phase.

The 40 f_h (low-pass carrier wave) produced in this way is fed to the sub-converter (at pin 2 of IC204), frequency-converted by 3.58-MHz from the 3.58-MHz VCO, and its sum component (4.21 MHz) taken out by the 4.21 bandpass filter. This output is fed to the main converter (at pin 8 of IC205).

Certain sections of the AFC system shown in Fig. 4-36 will now be described in greater detail.

Horizontal Synchronization Gate and Phase Detector

The horizontal sync gate opens approximately 16 μsec before horizontal sync and closes immediately after it. This action eliminates the equalizing pulse and vertical sync, whose phase differs from horizontal sync by 180 deg. Noise components within 40 μsec from horizontal sync are also removed. The horizontal sync signal, from which unnecessary components have been removed, is fed to the phase detector. The comparison signal comes out of the saw-tooth wave-generation filter as a saw-tooth wave at pin 14 of IC204. Here, the saw-tooth wave level during the horizontal sync period is detected, and the error voltage corresponding to the phase difference is fed to the VCO.

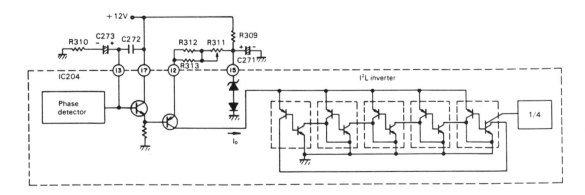

Fig. 4-37 VCO schematic diagram

160-f$_h$ VCO

The 160-f$_h$ VCO is a ring oscillator with inverters connected at odd stages (five stages in this IC) in a ring (see Fig. 4-37). The logic of this inverter integrated circuit is I^{2}L (Integrated Injection Logic). The first-stage I^{2}L inverter is composed of an npn transistor, which operates as an inverter, and a pnp transistor, which supplies base current to the npn transistor. The pnp transistor can be considered to be the collector load of the npn transistor in the previous stage. In other words, when the npn transistor is opened, all the current supplied from the collector of the pnp transistor becomes base current of the npn transistor of that stage, and this npn transistor becomes conductive to saturation. When the npn transistor in the former stage becomes conductive, all the current supplied from the collector of the pnp transistor is absorbed by the npn transistor in the previous stage. Base current is not supplied to the npn transistor, and it is, in effect, open. The time required for the npn transistor to change from Off to On gets faster as more current is supplied from the pnp transistor. The saturation—On mode—is released more quickly as the current absorbed by the npn transistor in the previous stage is greater. The collector current of this npn transistor increases as its base current increases. Thus, both On → Off and Off→ On operations are speeded up as the current supplied from the pnp transistor increases.

From the formula,

$$f = \frac{1}{2nTpd}$$

where

f = oscillation frequency of ring oscillator
Tpd = transmission delay time per inverter stage,

the oscillation frequency is seen to become lower as the transmission delay time increases. In the ring oscillator using an I^{2}L inverter, the oscillation frequency can be controlled by the current supplied from the pnp transistor, that is, by the current supplied from the emitter of the pnp transistor. Thus, the oscillation frequency is determined by the value of current supplied from the pnp transistor to the pnp transistor of every I^{2}L inverter, and consequently the self-oscillation frequency of the VCO can be adjusted by controlling this current with R311.

Four-Phase Rotation

Four-phase rotation is the center of the AFC system. Its output signal is the 40-f$_h$ low-pass carrier wave.

The phase is advanced by 90 deg every 1 H for the channel-1 side and delayed by 90 deg for every 1 H for the channel-2 side. The 160-f$_h$ VCO output is divided to ¼ by the ¼ divider in such a way that four 40-f$_h$ signals with phases different from each other by 90 deg can be obtained. At the same time, after f$_h$ has been divided by ¹⁄₄₀, it is further divided by ¼ to obtain ½-f$_h$ and ¼-f$_h$ signals. By combining these ½-f$_h$ and ¼-f$_h$ signals, the

During RECORDING

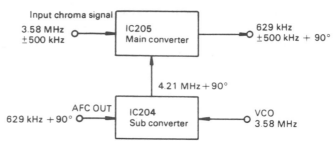

During PLAYBACK

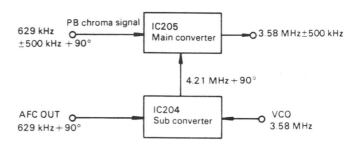

Fig. 4-38 Four-phase color rotation in recording and playback modes

above-mentioned four 40-f_h signals are changed over every 1 H to vary the phase by 90 deg. The phase in channel-1 is advanced 90 deg and in channel-2 is delayed 90 deg by means of the high or low of the SW30-Hz signal from the head drum tach. The "SW30-Hz signal" is the 30-Hz switching signal that controls the switching of the heads in such a way that channel-1 is advanced 90 deg and channel-2 is delayed 90 deg. The "SW30-Hz signal" is derived from the drum tach pulse and the 30-Hz reference signal, as shown in Fig. 5-10. The drum tach pulse signals are also called "30-PG signals," where PG is an abbreviation for pulse generator.

Thus, the low-pass converted chroma signal in the field recorded by the channel-1 head is recorded on the tape with its phase advanced 90 deg per 1 H. During playback, when the field played back by the channel-1 head is to be converted to the original 3.58-MHz signal, it is converted so as to remove the difference between the 4.21-MHz signal with its phase advanced 90 deg and the low-pass converted chroma signal with its phase advanced by 90 deg. Thus, the 90-deg phase advance is corrected, and the original chroma signal with continuous phase can be obtained. The color rotation treatment can be seen in Fig. 4-38.

ID Phase Inversion

Up to this point, the AFC system has been considered in the normal condition. However, phase disturbances may be caused when the switching point for the channel changeover drifts and when the rotation speeds in

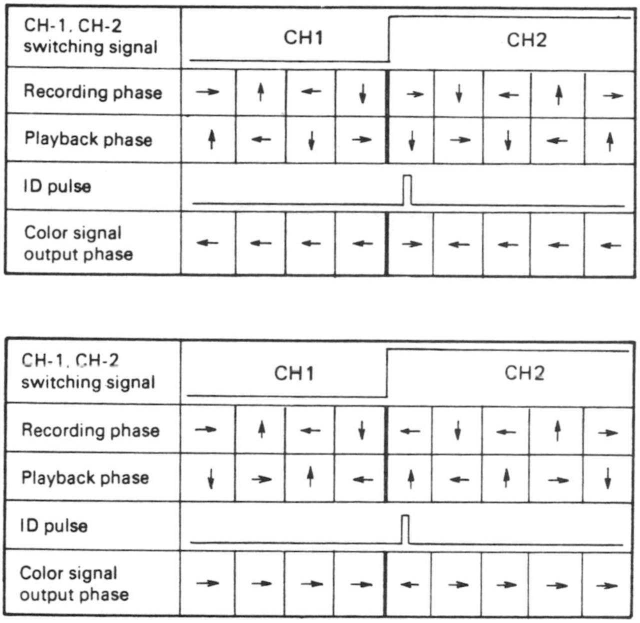

CH-1, CH-2 switching signal	CH1				CH2				
Recording phase	→	↑	←	↓	→	↓	←	↑	→
Playback phase	↑	←	↓	→	↓	→	↓	←	↑
ID pulse					⊓				
Color signal output phase	←	←	←	←	→	←	←	←	←

CH-1, CH-2 switching signal	CH1				CH2				
Recording phase	→	↑	←	↓	←	↓	←	↑	→
Playback phase	↓	→	↑	←	↑	←	↑	→	↓
ID pulse					⊓				
Color signal output phase	→	→	→	→	←	→	→	→	→

Fig. 4-39 Timing chart of compensation by ID pulse

video recording and playback differ. Servo timing errors can cause the video head switching to drift from the correct point. Such a case can occur if the servos lock in the wrong phase relative to the reference signal. Rotation speeds can easily be different if the recording is made on one machine and played back on another. Differences in back tension on the tape between the machines can easily cause this effect. As mentioned, ID pulse inversion functions to overcome the ill effects of these phase disturbances. They are not caused by dropout or the lack of synchronizing signal because the rotation signal is produced on the basis of f_h, which is obtained by dividing the VCO output.

When playback is started at a different phase from that in the recording, such as $\pi/2$ or $3/2\,\pi$, the output chroma signal at the point when channel-1 is changed to channel-2 or where channel-2 is changed to channel-1 becomes the output with a phase difference of π from the previous phase. When the phase differs by π, the ID pulse fed from the APC circuit inverts the 40-f_h phase by the phase inversion circuit and returns it to the same phase as that of the original field. When the switching point of the channel selector drifts, the same operation is performed. As described above in the case of phase inversion, the APC circuit cannot absorb this quickly enough. Thus, this circuit functions to invert the phase, to minimize the phase drift that is absorbed by the APC circuit, and to reduce the burden of the APC circuit.

Figure 4-39 shows timing compensation by the ID pulse.

5

Servo System Operation

The purposes of servo systems in video recorders are as follows:

1. To run the magnetic tape at a constant speed of 3.335 cm/sec in the SP (standard play) mode and at 1.667 cm/sec in the LP (long play) mode

2. To rotate the video head at one-half the vertical sync frequency of the input video signal and to set the relative positions of the video head and the tape so that the vertical blanking period can be recorded on the lower end of the tape video track. In other words, the joint or switching point of the picture is fixed at the lower part of the screen during playback. Figure 5-1 shows the VHS magnetic tape pattern.

In general in the 1 field/track system of a two-head video tape recorder for home use, the video signal fields overlap at the switching point. The system is designed so that the lack of signal at the joint or switching point is removed during playback and that noise at the joint is not conspicuous by fixing the joint at the lower part of the picture.

The purposes of the playback servo are (1) to run the magnetic tape at a constant speed, the same as that in recording, and (2) to rotate the video head precisely (30 Hz) in order to trace the recorded video tape track accurately—that is, to perform tracking. Vertical blanking follows in the

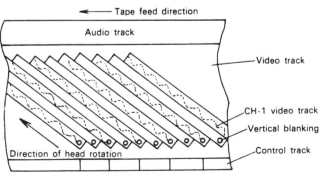

Fig. 5-1 VHS magnetic tape pattern

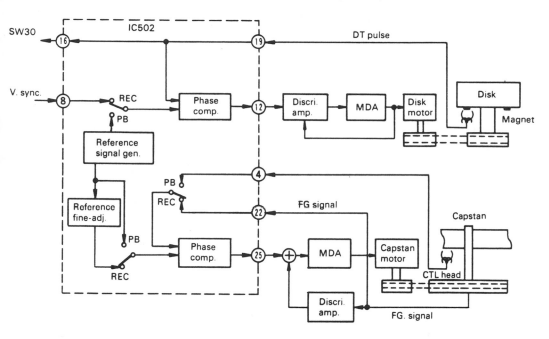

Fig. 5-2 Overall diagram of drum (disk) and capstan servo systems

retrace from the bottom to the top of the picture, and about 35 lines of any NTSC signal is blanked out during that period.

When the head rotation is uneven, the length 1 H on the track becomes uneven. The position of the H sync signal is recorded irregularly, and the length of the playback video signal varies every H in terms of H, producing a picture with a great deal of jitter. Damping action is provided by the servo system to prevent these phenomena.

The servo circuits consist of the capstan servo, drum servo, and the reference signal generator (see Fig. 5-2). The capstan servo synchronizes the REF30 (30-Hz reference) signal and FG (frequency generator) signal, which is detected from the capstan base during recording, and synchronizes the REF30 signal and the control pulse, which is played back from the tape during playback. The drum servo synchronizes the vertical sync signal and the drum tach pulse during recording and synchronizes the REF30 signal and the drum tach pulse during playback.

The Capstan Servo System

The capstan servo system is designed to ensure that the magnetic tape runs at a constant speed of 3.335 cm/sec in the SP mode and at 1.667 cm/sec in the LP mode during both recording and playback.

Since a dc motor is used in this design, two servo systems are required. Some VCRs use ac motors, which can be handled with one capstan servo system. The rotation speed of dc motors can change as a result of load fluctuation as well as when the voltage applied to the motor changes. In this design, one control system is called the discriminator-

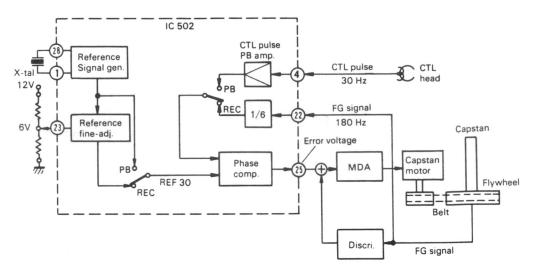

Fig. 5-3 Capstan servo system

control system, and the other is called the phase-control system. The discriminator-control system controls large speed variations to achieve a constant speed. The phase-control system controls small variations to achieve the specified tape speed.

Capstan Servo System Operating Principles

The operating principles of the capstan servo system are shown in Fig. 5-3. The reference signal in the phase-control system in the capstan servo is a crystal oscillator with an oscillation frequency of 32.768 kHz. This frequency is divided by 1093 in the internal counter of IC502 and is applied to the phase comparator as the 30-Hz reference signal.

For the comparison signal during recording, the 180-Hz FG signal is obtained in the SP mode (90-Hz in the LP mode) from the frequency generator, which utilizes a pair of magnets embedded in the capstan base, and fed through pin 22 of IC502; it is divided by 6 in the SP mode (3 in the LP mode) to become the other input of the phase comparator. During playback, on the other hand, the 30-Hz control pulse picked up from the tape is amplified and shaped by the control head to become the other input of the phase comparator. The phase comparator generates a voltage (the error voltage) that is proportional to the phase difference between the reference signal and the comparison signal. This error voltage is fed from pin 25 of IC502 to the output of the discriminator control system by means of IC501. It is further fed to the capstan motor via the motor drive amplifier (MDA).

When this phase-comparison control system is used alone, large frequency variations such as those during start-up cannot be controlled. Consequently, a discriminator control system is required. It is composed of the FG amplifier (Q514 to Q517) and the speed detector (Q512/Q513). This system suppresses frequency changes during start-up or at times of large load changes and achieves approximately the correct rotation speed.

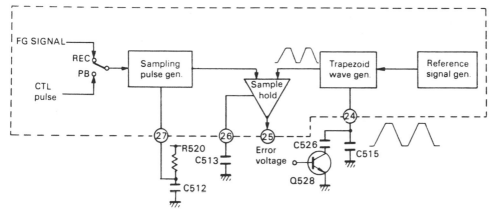

Fig. 5-4 Phase comparator

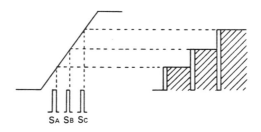

Fig. 5-5 Sample and hold

Reference Signal Generator

The crystal, which is inserted between pins 1 and 28 of IC502, oscillates at 32.768 kHz. Its output is counted down by 1093—by the counter inside the IC—to become 29.98 Hz (REF30). This REF30 is the reference signal for both the capstan and the drum servo during playback but only for the capstan servo during recording. The frequency of the reference signal can be changed in three steps for tape-speed adjustment by $\pm$ 0.3 percent with respect to the reference frequency, 29.98 Hz.

Phase Comparison

The circuit in Fig. 5-4 performs phase comparison between the reference signal and comparison signal and holds an error voltage. A trapezoid waveform is obtained at pin 24 of IC502 by means of REF30, which is input from the reference signal generator. At the same time, the sampling pulse is generated from the comparison signal in the pulse generator, and the rise-slope section of the trapezoid wave is sampled with this pulse.

The sampling voltage is held with capacitor C513 at pin 26 of IC502 and is fed out through pin 25 as the error voltage. As shown in Fig. 5-5, the sampling hold output falls when the comparison signal (sampling pulse) has advanced (as at S_a) compared with the reference signal (the trapezoid wave), and the sampling hold output rises when the comparison signal is delayed (as at S_c). For example, during recording, when the capstan speed decreases, the error voltage becomes higher.

Discriminator Control

As mentioned previously, the discriminator control system is required to provide damping at start-up time or when large fluctuations occur. Functioning as a frequency discriminator, it converts frequency changes to voltage changes. The discriminator control circuit in the capstan servo is shown in Fig. 5-6. Its operating principle in the SP mode is described below.

The FG signal is fed in and fully amplified by the FG amplifier consisting of Q515 to Q517, in order to drive Q514. Assuming that Q514 changes from On to Off, the Q514 collector potential V1 changes from low to high, and the Q513 emitter potential V2 fed via capacitor C519 becomes high simultaneously. In fact, V2 becomes higher than the power voltage at that time, and Q513 becomes conductive. Then, the potential of C519 charges C518 via Q513, and the output voltage, Vout, is increased during this period. When C519 continues to discharge for a certain time, V2 falls and Q513 is opened. As a result, C518 starts discharging via R523 and R524, and Vout gradually falls. When Q514 changes from On to Off again, Vout increases in the same way, and the average potential settles to a constant value.

Assuming that the capstan speed increases and that the frequency of the FG signal is raised, the charging time of C518 is fixed but its discharging time is shortened. As a result, the average value of Vout is raised. Accordingly, the output at pin 1 of IC501 falls, and the capstan motor is slowed to rotate at the specified speed.

When 12 V is now applied to the base of Q512 through R617, that transistor becomes conductive, thereby adding dc bias and making the speed of operation in the LP mode one-half that in the SP mode.

Fine Adjustment of Tape Speed

During recording, the capstan servo controls the capstan shaft. Slight speed fluctuations can occur as a result of the shaft tolerance. To compensate for these fluctuations electrically, a REF fine-adjustment control is provided. By changing the frequency of REF30 slightly, the tape speed is stabilized within the specified range. In this design, REF30 is fine-adjusted by changing the step-down ratio of the reference signal generator by varying the voltage at pin 23 of IC502. By changing the voltage at pin 23 to 12 V, 6 V, or 0 V, tape speed can be adjusted by ± 0.3 percent (0 percent at 6 V).

Capstan Motor Control

In Fig. 5-7, when low voltage is applied to Q521 from the system-control circuit, Q521 is opened. By controlling the conductivity of Q521, the capstan motor may be stopped or started. When the output from pin 1 of IC501 is fed to the motor drive amplifier (MDA-Q522/Q523), the capstan

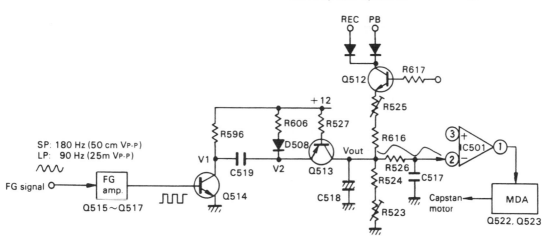

Fig. 5-6 Capstan discriminator control circuit

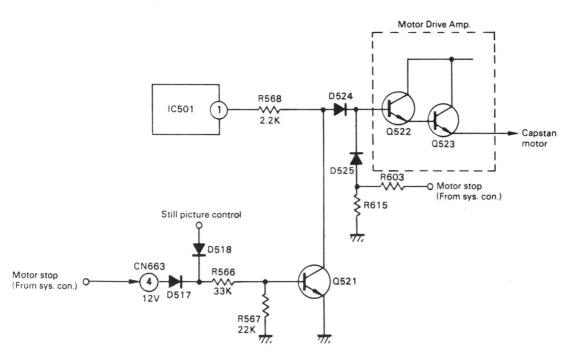

Fig. 5-7 Capstan motor control circuit

motor normally starts rotating. However, when high voltage is applied from the system-control circuit, Q521 becomes conductive. Accordingly, the collector of Q521 becomes approximately 0 V, the output at pin 1 of IC501 becomes 0 V at the input side of the MDA, and the capstan motor stops. When the unit enters the remote pause mode during playback, the high voltage is applied to D518, and Q521 becomes conductive. This action stops the motor, and still pictures are played back.

Switching of Error Voltage

The presence or absence of the control pulse is detected, and the error voltage is switched in the playback mode. When the control pulse is not present at pin 2 of IC502, Q508 is always open, C509 is charged, and Q509 becomes conductive. At that time, Q510 becomes conductive and saturated, and both its collector and emitter are at approximately the same potential. Accordingly, the gate voltage of FET Q511 is fixed at a constant voltage, and the error voltage from pin 25 of IC502 is not received. When the control pulse is delivered, all the above-mentioned transistor modes are reversed. When Q510 becomes open and the error voltage at pin 25 of IC502 is applied to the gate of FET Q511, the servo control operates normally. Rise of the motor speed is improved during start-up by detecting the control pulse and during playback by switching the error voltage. At the same time, the tape speed is not changed when playing back a blank tape.

Drum Servo System

The drum servo system (Fig. 5-8) controls the rotating video head drum (disk)—which carries the video heads—at a constant synchronized rotation speed (30 Hz) during both recording and playback. The control systems are composed of the phase control system and the discriminator control system, just as in the capstan servo system. The vertical synchronizing signal of the input video signal is used for the reference signal of

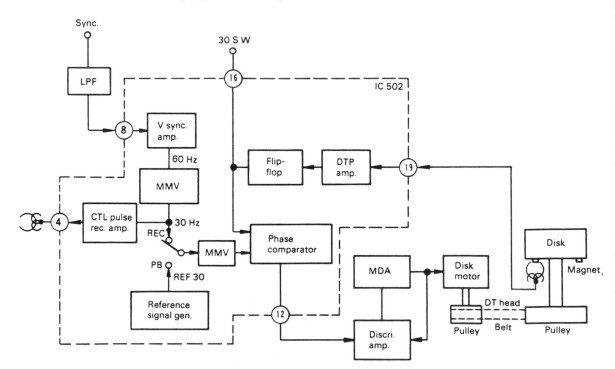

Fig. 5-8 Video head drum (disk) servo system

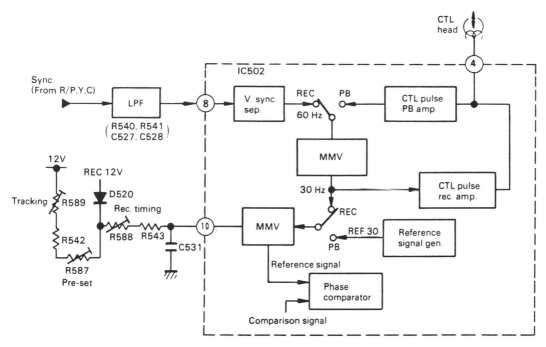

Fig. 5-9 Reference signal diagram

the phase control system during recording. The 30-Hz signal divided from the 32.768-kHz crystal-oscillator signal is used for the reference signal during playback. At the same time, the comparison signal detects the rotation phase from the drum tach head and compares the phase with these two reference signals in order to control the drum rotation. The discriminator control system smooths any unevenness in the motor speed during start-up or when the load changes and applies damping to keep the rotation speed constant.

Drum Servo System Operating Principles

As mentioned above, the reference signal to be used in the phase control system of the drum servo system depends on whether the mode is recording or playback. During recording, the vertical sync signal, which is contained in the video signal, is fed as the reference signal to the phase comparator. During playback, REF30 (the 30-Hz signal developed by the reference signal generator) is used.

In the phase control system of the recording drum servo system, the sync signal is fed in from the R/P,Y,C printed circuit board as the reference. This signal becomes the complete vertical sync signal (60 Hz) at the vertical sync separator via the LPF in the upper left-hand corner of Fig. 5-8 because of the servo circuit. It is divided by 2 in the counter monostable multivibrator in the next stage to become 30 Hz. Part of this signal is applied to the phase comparator via the monostable multivibrator and part to the control pulse recording amplifier. The comparison signal is

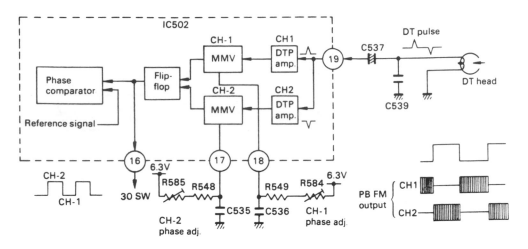

Fig. 5-10 Drum (disk) tach pulse (DT) route

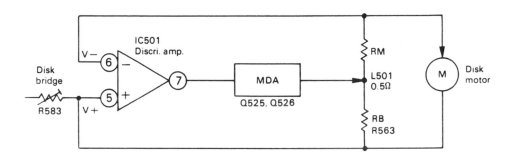

Fig. 5-11 Drum (disk) discriminator circuit

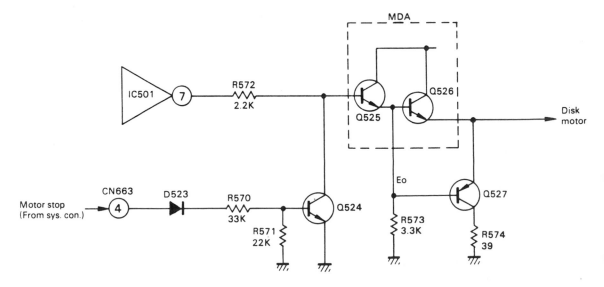

Fig. 5-12 Drum (disk) motor control circuit

detected from the two magnets embedded in the video head drum beside the DT (Drum Tach) head. It is fed to the DTP amplifier, which amplifies and shapes it. Then it is converted to 30 Hz via the flip-flop to be applied to the phase comparator.

The phase comparator converts the comparison signal to a trapezoid wave, changes the reference signal to a sampling pulse, and generates a sampling error voltage corresponding to the synchronized phase at that time. This phase-error voltage is applied to the discriminator amplifier and performs the damping control needed for the large speed changes that occur during start-up or are caused by external disturbances.

By means of the switching shown in Fig. 5-8, the REF signal of the reference signal generator is selected as the reference in the drum servo system for the playback mode. The comparison signal becomes the drum tach pulse as it is in the recording mode, and the servo loop is entirely the same as in the recording mode.

Reference Signal Route

The sync signal input from the R/P,Y,C printed circuit board is applied to pin 8 of IC502 after its vertical sync component has been taken out by the low-pass filter (LPF) shown in the upper left-hand corner of Fig. 5-9. Inside IC502, the vertical sync signal is shaped and divided by 2 by the counter monostable-multivibrator in the next stage, thus becoming 30 Hz. This 30-Hz signal is fed to switch 2 and at the same time fed to the control head via the control pulse recording amplifier.

The counter monostable-multivibrator determines the pulse width by counting a certain optimal number of crystal oscillation signals. This action differs from that of a conventional circuit, which uses the RC time constant. By making this pulse width a little larger than the vertical sync, the vertical sync signal is divided by 2, making it 30 Hz. The output of the counter monostable-multivibrator is applied with switch 2 to the delay monostable-multivibrator in the next stage during recording. The delay time of this monostable-multivibrator is determined by the parts externally attached to pin 19 of IC502.

Since + 12 V is applied to the anode of D520 during recording, the delay time in that mode is determined by C531, R543, and R588. R588 is the video-recording timing-adjusting variable resistor. Its purpose is to set the recording of the video vertical sync signal at a specifed position on the tape. During playback, the + 12 V is not supplied to the anode of D520 when the REF30 signal is applied to the monostable-multivibrator. Then the delay time is determined by C531, R543, and R588 and the combination of R542, R587, and R589. The delay time during playback can be much longer than during recording. R587 and R589 become the tracking adjusters.

Drum Tach Pulse Route

The drum tach pulse, which is the source of the comparison signal in the phase control system, is obtained by detecting the rotation of the

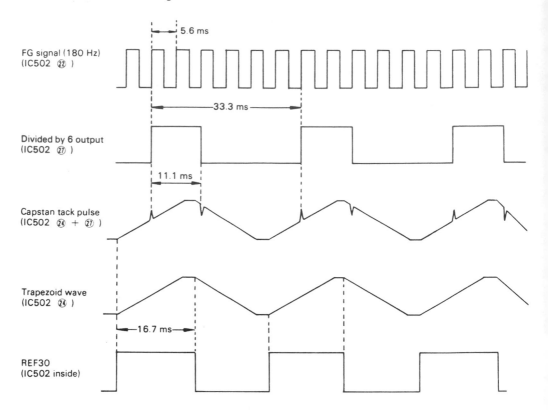

Fig. 5-13 Timing chart of the recording capstan servo diagram

magnets that are embedded in the drum at positions 180 deg from one another with different magnetic polarities. The drum pulse is applied to pin 19 of IC502, and positive and negative pulses are amplified and shaped separately by the DTP (Drum Tach Pulse) amplifier (see Fig. 5-10). Both these positive and negative pulses trigger the phase-adjusting monostable-multivibrator. By changing the delay time of this monostable-multivibrator by means of the externally attached R584 and R585, the phase of the output of the channel-1 and channel-2 video heads can be changed (the output at pin 16 of IC502).

Discriminator Control

As shown in Fig. 5-11, the discriminator control system of the drum is very different from that of the capstan servo. Output voltage from pin 12 of IC502 is fed to V+ (pin 5 of the discriminator amplifier, IC501) via Q518 and Q519. At the same time, the MDA (Motor Drive Amplifier) output is fed to V− (pin #6 of IC501) via Rm. The potential difference of these voltages is amplified by the discriminator amplifier, impedance-converted by the MDA, and fed to the drum motor via Rm (drum motor control resistor). In the operation of this circuit, when the motor speed starts falling, the current flowing into the motor via Rm becomes larger; consequently, the potential of V− falls. These changes are detected by the

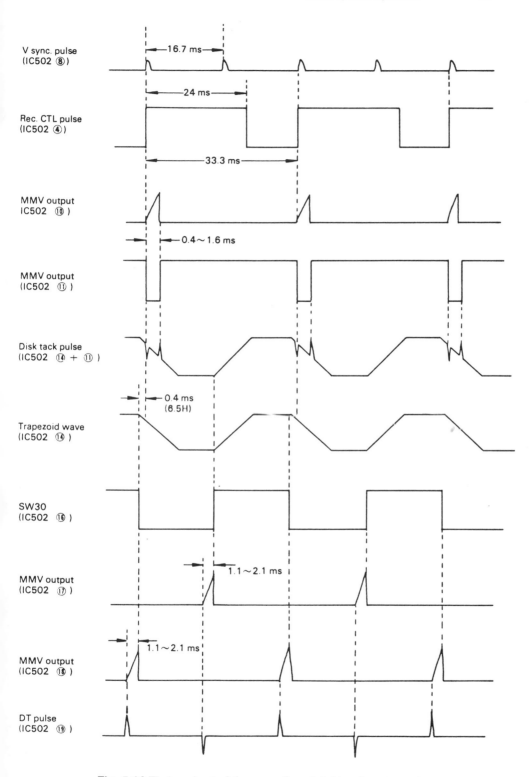

Fig. 5-14 Timing chart of the recording disk (drum) servo system

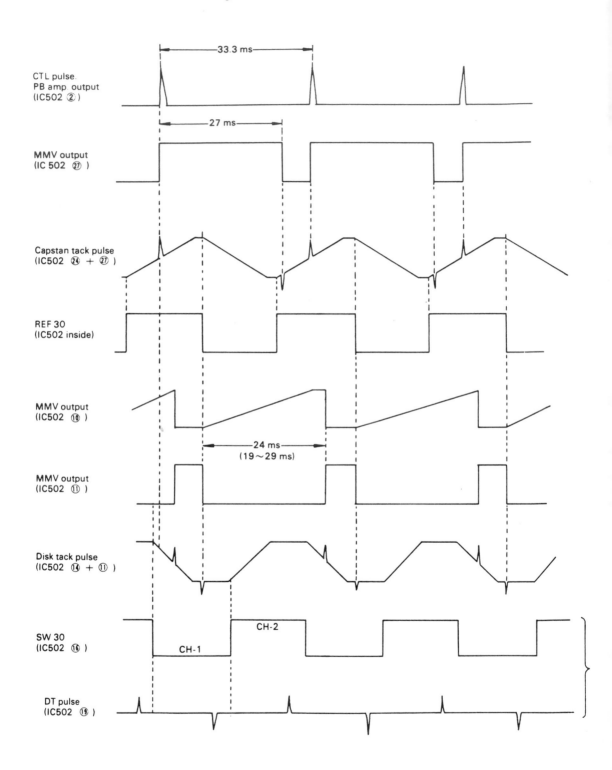

Fig. 5-15 Timing chart of the playback servo system

discriminator amplifier, and a higher voltage is generated at its output. This voltage is applied to the motor via the MDA, and the motor speed is increased and, by such action, kept constant.

Drum Motor Control

In Fig. 5-12 the MDA (Q525/Q526) and Q524 operate as they do in the motor control of the capstan servo. In other words, when low voltage is fed from the system control circuit, the drum motor starts, and when high voltage is delivered, the drum motor does not rotate.

The purpose of Q527 is to speed up the lock-in when the drum motor abruptly enters the no-load mode for some reason, and the load is restored again. When the drum motor is rotating with normal load, Q527 is opened, and the emitter voltage of Q526 is as follows: $E_0 - V_{bn}$ (npn) $= E_0 - 0.7$ V. When the motor mode becomes no-load, the emitter voltage of Q526 rises because of the counter electromotive force of the motor, and when $E_0 + V_{be}$ (pnp) becomes more than $E_0 + 0.3$ V, Q527 becomes conductive and the counter electromotive force of the motor is grounded. As a result, the motor speed falls, Q527 opens again, and the motor returns rapidly to the lock-in mode.

Figures 5-13, 5-14, and 5-15 show timing charts of the servo system.

6

System Control Operation

System Control Circuit

The operations of the system control circuit can be roughly classified as follows:

1. *Operation lever reset (auto shut-off):* When it is desired to stop operation of the mechanism at the end of the tape or when abnormal conditions occur during the running of the tape, such as when the drum rotation stops, the stop solenoid of the control circuit automatically resets the operation controls to the stop mode.
2. *Motor operation instructions:* The system control circuit gives instructions for the rotation of the drum and capstan motors.
3. *Loading motor instructions:* The system control circuit gives instructions for rotation of the loading motor to perform loading and unloading or remote-control pause operations.

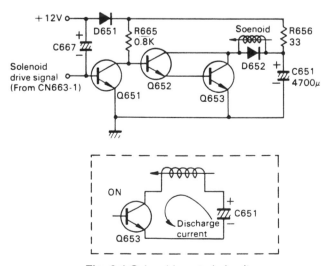

Fig. 6-1 Solenoid control circuit

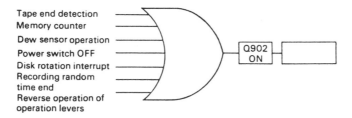

Fig. 6-2 Automatic shut-off diagram

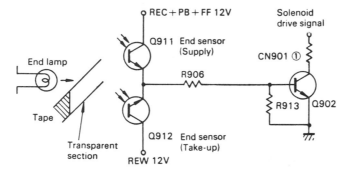

Fig. 6-3 End sensor circuit

4. *Video recording start timing and monitor cut:* The system control circuit controls the tape recording timing and the timing of the picture (monitor cut) on the TV screen during playback.

5. *Holding power supply:* The system control circuit gives instructions for the start and hold of the 12-V power supply.

Operation Lever Reset (Auto Shut-Off)

The diagrams for operation lever reset are given in Figs. 6-1 and 6-2. Voltage H is always applied to the base of Q651 when the tape is traveling normally, and the circuit is such that this voltage becomes "L" during any abnormal condition. Q651, Q652, and Q653 are on the servo printed circuit board, and the voltage applied from the system control printed circuit board is fed to their bases by the solenoid drive signal. When this signal becomes "L," Q651 opens, and the base voltage of Darlington-connected Q652 and Q653 becomes "H," making them conductive. Energy charged in C651 is discharged via the solenoid and Q653. This solenoid is called the *operation-lever-reset solenoid.* When current flows, it releases the lock mechanism that locks the operation levers, and the video deck enters the stop mode (H = high; L = low).

The base voltage of Q651 becomes "L" when Q902 becomes conductive. The conditions for Q902 becoming conductive are as follows:

1. *Tape-end detection:* When the tape is at its start or end.

2. *Memory counter:* When the memory counter function operates during rewinding.

3. *Dew sensor operation:* When there is a signal from the condensation or dew detector.
4. *Drum rotation interrupt or stopper:* When the drum rotation stops while the tape is traveling.
5. *Power switch off:* When the power switch is turned Off while the tape is traveling.
6. *Reverse operation of operation levers:* When the operation levers are operated in reverse (when levers giving the operation instruction are lifted).
7. *Video recording random end:* When the Off instruction is given from the end interval timer during the timer video recording.

Tape End Detector or Sensor (Q911/Q912)

Details of the circuit for the tape end detector are shown in Fig. 6-3. When the tape comes to its start or end during running, it is detected and the video deck must enter the stop mode. A transparent section of approximately 20 cm is provided at both ends of the video tape for this purpose. The end detector lamp (on the timer printed circuit board), the end-detecting phototransistor Q911 (on the supply side of the end-sensor printed circuit board), and Q912 (on the takeup side of the end-sensor printed circuit board) detect the above condition optically. During video recording, playback, and fast-forward, + 12 V is applied to the collector of Q911. When the light falls on Q911 through the transparent section at the end of the tape, the transistor becomes conductive, and current flows through R906 to make Q902 conductive. Accordingly, the solenoid control signal becomes "L."

The rewind signal (12 V) is applied to the collector of Q912 during rewinding, Q912 becomes conductive at the rewind side tape end, Q902 becomes conductive, and the solenoid control signal becomes "L." R913 is a resistor that determines the sensitivity of photo-transistors Q911 and Q912.

Memory Counter

The counter switch sets the standard for the start position of video recording. The memory counter operates to set the counter switch to "000" at the start of video recording and to stop at "999" during rewind (since "999" is next to "000" during rewind, it can be considered to stop at approximately the same position). When memory switch S907 is set On during rewind, counter SW906 turns On at "999", and the base of Q902 is thus set to "H" via the differential circuit C,R. As a result, the solenoid control signal becomes "L," and the operation lever is reset by the solenoid.

Dew Sensor Operation (Q101 and Q102)

When dew forms inside the video deck, obstacles to the running of the tape include adherence, breakage, or tangling of the tape. The dew sensor detects the condition and avoids accidents stemming from it. When condensation occurs, it is necessary to stop the tape. The dew sensor and Q101 on the timer printed circuit board perform the necessary detection

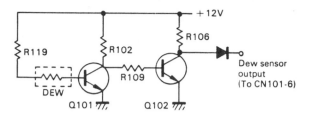

Fig. 6-4 Dew sensor circuit

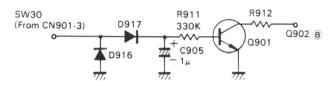

Fig. 6-5 Disk (drum) rotation stopper

(Fig. 6-4). Q102 inverts their signal and sets the base of Q902 to "H" via connectors CN101/CN106. As described above, when the base of Q902 is set to "H," the operation buttons are reset.

Drum Rotation Stopper (Q901)

It is necessary to stop the tape from running to prevent damage to the drum motor or any other damage occurring from an accident in which drum rotation is inhibited by an abnormal condition such as dew binding the tape to the drum. SW30, which is fed to pin 3 of CN901 from the servo printed circuit board, is the 30-Hz pulse signal during drum rotation (see Fig. 6-5). It is clamped by D916 and integrated by C905 and R911. Signal "H" is applied to the base of Q901 during the feeding of the SW30 signal. Q901 is conductive during recording and playback, and only the "L" signal is applied to Q902. When the drum stops, the SW30 signal disappears, and the base voltage of Q901 becomes "L." Accordingly, Q901 opens, the base of Q902 becomes "H" via Q901 and R912, and the operation buttons are reset.

Power Switch Off Operation

When the power switch is set to Off during tape running, the heater 18 V from the timer printed circuit board becomes "H," thus setting the base of Q902 to "H" via R917, R906 and D915. As a result, the operation buttons are reset.

Reverse Operation of Operation Levers

In this recorder, the operation levers, which give instructions during tape running, can easily be turned in the reverse direction (to the stop position). At this time, despite the tape's being in the running mode mechanically speaking, it is in the stop mode so far as the signal is concerned, which causes somewhat loose tape. Accordingly, it is necessary to reset the operation levers during the reverse operation to stop the tape mechanically.

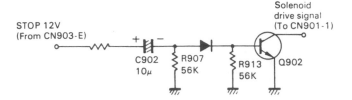

Fig. 6-6 Stop 12-V signal to system control PCB

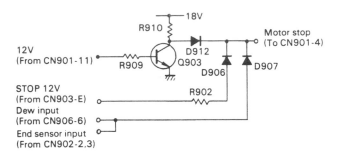

Fig. 6-7 Motor stop circuit

In the stop mode, the stop 12-V signal is sent to the system control printed circuit board from CN903-E (see Fig. 6-6). This signal is applied to the base of Q902 via the differential circuit of C902 and R907, which makes it conductive instantaneously. As a result, the solenoid control signal becomes "L," the solenoid is activated, and the operation levers are reset. (Without the stop lever's being pressed, for example, the stop mode is entered if the record levers are pressed in reverse during recording.)

Recording Random End

The random timer gives the recording-time end signal during timer recording. This operation is described separately, but it is necessary to reset the operation levers to stop the tape from running when this end instruction is given. Called the *end out signal*, it is sent to the timer printed circuit board from the random timer printed circuit board. Then, together with the dew detection signal, it is applied via D111 to set the base of Q902 to "H" and reset the operation levers.

Motor Operation Instructions (Capstan and Drum)

The system control circuit gives only rotation instructions for the capstan and drum motors. Speed control for these motors is provided by the servo circuits.

The motor stop rotation instruction (Motor Stop) is sent to the servo printed circuit board via connector CN901-4 (see Fig. 6-7). There it instructs motor stop action when it is "H" and motor rotation when it is "L." When the power switch is Off, Q903 is 18 V, and the "H" stop signal is applied via R901 and D912. When the power switch is turned On, 12 V sets the base of Q903 to "H" via R909, thereby making Q903 conductive. When

12 V is applied under the stop mode, via R902 and D906, this stop 12 V is set to "H" in the same way. However, when one of operation switches S901/ S904 is pressed, this stop 12 V is set to "L," and both motors rotate. The Unload 2 signal from the loading motor PCB is "H" during unloading and stops both motors. At the same time, the Unload 1 signal is applied to the capstan motor via the loading motor PCB, the system control PCB, and the servo PCB and gradually rotates it so that unloading will not damage the tape.

During dew detection and tape end detection, both of which require solenoid operation, this signal is applied via D907 to set the Motor Stop signal to "H" and inhibit tape running.

Loading Motor Operation Instructions

Figure 6-8 shows the operation circuit of the loading motor drive system and the pause system that uses it. In the stop mode, loading switch S931 is on the ground side, and since the unloading switch is grounded via Q932, the loading motor does not rotate. When the record or play lever is pressed, S931 is switched to the 12-V power supply side mechanically; current flows via S931, the loading motor, S932, and Q932; and the motor rotates normally to enter the loading mode. When loading is completed, S931 is switched to the ground side mechanically, and the stop mode is entered again. S932 is also switched mechanically to the 12-V power side during unloading. Then current flows to S932, the loading motor, and S931, and the motor reverses and performs unloading. When unloading is complete, S932 returns to its original position and enters the stop mode.

In the pause mode—for both remote-control pause and mechanical pause—the pause signal is sent to the loading motor PCB and the system control PCB. This pause signal makes Q931 conductive; Q935 connected to the collector side opens; Q933 becomes conductive; and 12 V is supplied in the reverse direction to the loading motor. Q932, opened by means of Q931,

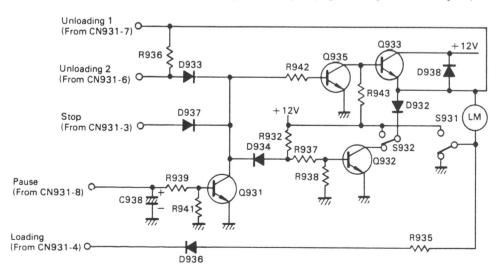

Fig. 6-8 Loading motor operation circuit

is not short-circuited by the sequence, of Q933→S932→Q932. Accordingly, current flows in the sequence, Q933→loading motor→S931. However, S931 is mechanically switched to the 12-V side immediately after unloading is performed during recording or playback so that the voltage at both ends of the loading motor is changed to "H." The tape stops running under this mode—the unload mode—so that the compression of the pressure roller can be released.

In the case of mechanical pause, compression of the pressure roller is also released mechanically, and the pause is made possible by applying a light force to reverse the loading motor.

Video Recording Start Timing, Monitor Cut

To record the video signal, audio signal, and control signal with the same timing during recording, it is essential to make a good connection between the previous video signal and the new video signal. For this purpose, the loading signal ("H" during the loading period) from the loading motor PCB is charged to C904 for some delay, and the recording is started as soon as tape running is stabilized—approximately 1 sec after loading is completed. This signal that has delayed the loading signal is called the *recording start signal*. It is sent to the servo PCB from the system PCB and is then fed to the video system. It would be the same during playback, but the design is such that the circuit does not operate in that mode.

The *monitor cut* functions to prevent distorted pictures from appearing on the TV screen during playback. This signal, which is similar to the recording start signal, has the pause signal added to it on the servo PCB. Nothing appears (during the blanking mode) on the TV screen for approximately 3 sec after the play lever is pressed, and the blanking mode is entered during pause. Q934 prevents the loading signal from entering C904 during pause (to prevent skipping when the pause lever is pressed and released in the recording mode). It is designed so that C904 will be charged from stop 12 V via D914 to prevent white noise from appearing on the TV screen as a result of the delay in switching S931.

Power Hold

It is necessary to feed a signal to the power stabilizer circuit (servo PCB) in order to generate and hold the 12-V power supply.

When S102 (the power switch) is turned On while S101 (the timer switch) is Off, the 18-V power hold signal is fed to the servo PCB via the timer system control PCB to generate 12 V. The record or the playback 12 V is applied to this signal line via D903/D904. It is designed so that power will be maintained by means of the Unload 2 signal until unloading is completed, thus enabling the tape to be taken out when S102 (the power switch) is turned Off during tape running. The record 12 V or playback 12 V functions to maintain the 12-V signal (power supply) for the purposes of the *sleep operation* (to be explained in the description of the sleep function) and the timer recording operation. The timer output is differentiated to start the power-stabilized circuit and generate 12 V during the timer recording operation, and the 12 V is held by the record 12 V after that.

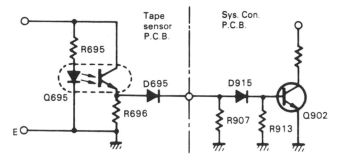

Fig. 6-9 End-of-tape sensor circuit

Tape Sensor Operation

When the torque of the take-up reel drum decreases, the tape may drop out of the tape running path and create the abnormal condition of jammed tape. The purpose of the tape sensor is to prevent this from happening. The tape sensor (Q695) is composed of a photo transistor and an LED (emitting infrared rays) lined up in parallel (see Fig. 6-9). When the tape runs near the sensor, the photo transistor detects the infrared rays, and a detection signal is produced. This signal drives a solenoid, which releases operation levers to stop the tape.

The tape sensor is installed on the removable plate (for opening/closing the tape protection lid) near the capstan and pressure roller. Its function is to make sure that the tape is running correctly at all times.

When the tape runs near the tape sensor, the photo transistor in Q695 receives the reflected rays from the tape and conducts. Current flows to Q902 via D695/D915, and Q902 becomes conductive. As a result, the potential of the solenoid drive signal becomes "L," the solenoid is drawn in, and the operation levers are reset.

The Timer Circuit

The functions of the timer circuit are the following:

1. Detection and indication of dew
2. Timer recording operation and random timer operation
3. Generation of a 2-min pulse for random timing
4. Sleep operation

Dew Indication

The operation to detect dew and reset the operation levers was described in the discussion of the system control circuit. Here we will describe the indication circuit shown in Fig. 6-10. Q103 changes the power indicator LED, D106, from lit to flashing when dew occurs. When there is no dew, Q101 opens, Q103 is conductive, and D106 lights. When condensation occurs and Q101 becomes conductive, Q103 becomes conductive; it is opened every second by means of the 1-Hz output of the timer IC (IC101).

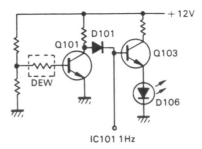

Fig. 6-10 Dew indication circuit

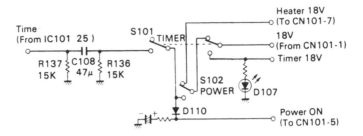

Fig. 6-11 Timer output circuit

The dew sensor itself is a variable resistance—high when dry and low when wet. It has quick-detection operation.

Timer/Random-Time Recording

During timer recording, S101 (the timer switch) is turned On (the switch is pushed) to enter the timer mode (see Fig. 6-11). S101 has precedence over S102 (the power switch), and the 12-V power supply is turned Off.

When the recording time and the present time become the same, the timer output is fed from IC101, and the differential signal is supplied to the stabilized power circuit of the servo PCB via differentiating circuit C108 and R136. When this occurs, the 12-V power supply is generated, held in the mode when the record or play lever is pressed, and tape running starts. When the operation lever is not pressed, 12 V is generated instantaneously and tape running is turned Off.

The random timer finishes the recording in accordance with the length of the recording program in the timer recording mode. The 2-min pulse from the timer is counted, and the timer recording is finished at the time set by the 15-min, 30-min, 60-min, and 120-min selector switches. (As described in the discussion of the system control circuit, the operation levers are reset to enter the stop mode. When unloading is completed, the signal that holds 12 V is terminated entirely, leaving the 12-V power Off and only 18 V remaining). By pushing the 15-min and 30-min selector switches, timer recording for 45 min is possible, and a maximum of 3 hr and 45 min is possible in 15-min intervals. When none of these selector switches are turned On, recording continues until the end of the tape.

2-Min Pulse Counter

By differentiating and adding the B and E (segment) output of the "MINS" (minute) output of IC101 and clamping to "H," it is possible to produce "L" pulses every 2 min. IC121 and LC122 count these 2-min pulses. By counting in binary, 8, 16, 32, and 64 times this amount of time can be obtained. This output drives the solenoid as the end out signal via S121/S124.

The Sleep Operation

When S101 (the timer switch) is turned On during recording or playback, the power supply of 18 V from S102 is cut, but the 12-V power is held by the signal from the record 12 V or the playback 12 V, and the running mode is continued. However, when the operation lever is released at the end of the tape and unloading is complete, the supply source of the power hold signal is removed and the 12-V power supply is turned Off. In this way, the sleep operation saves power during the night. The random timer previously described can be used as the sleep agent.

7

VTR Adjustment Tools and Test Equipment

Anybody who services color TV receivers has a fair start toward the electronic test equipment required for video recorder servicing. Additional test equipment is essential, however, and several tools for mechanical adjustments and tests must be acquired. Moreover, one needs test/ alignment tape cassettes for the various formats; these are available from the company that sells the recorder. The video recorder manufacturers rely heavily on the test/alignment tapes in their procedures for electronic adjustments. If one has invested in a versatile video generator (such as the Sencore VA48) for color TV service work, this instrument will suffice for the signal generation and circuit analysis needed for your VTR work. However, the VTR maker sometimes recommends an NTSC color video generator as the color TV signal source, but with alternatives—for example, the gated rainbow color pattern generator. Other essential test equipment includes a dual-trace oscilloscope of suitable frequency range, a frequency counter, and a digital multimeter. A moderately well-equipped color TV service shop may already have some of this equipment and can also be expected to offer items like transistor testers and standard meters.

Mechanical Adjustment Tools

For the mechanical servicing of their Betamax recorders, the Sony Corporation of America recommends and sells the tools and fixtures shown in Fig. 7-1. They include the following:

1. Semifixed coil adjustment tool
2. Forward back-tension measurement fixture
3. T-type and R-F type torque measurement cassettes
4. Tension regulator forward position fixture
5. DC motor gear spacer
6. Threading ring clearance gauge
7. Cassette reference plate

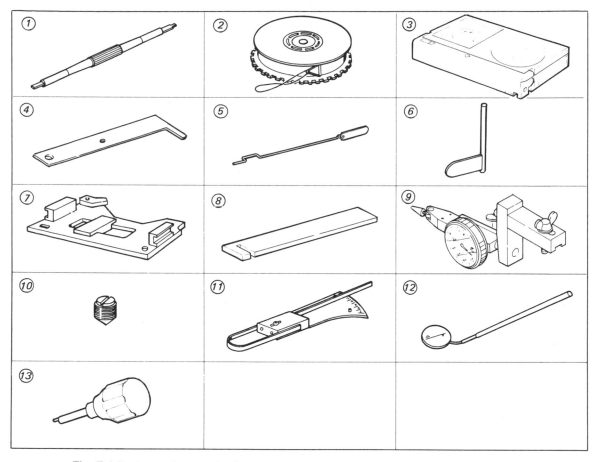

Fig. 7-1 Tools and fixtures for Betamax recorder maintenance (*Courtesy,* Sony Corp. of America)

8. Two tension regulator bending fixtures
9. Upper head drum alignment fixture (eccentricity gauge)
10. Four dihedral adjustment screws
11. Tension gauges (max. 100 g, 200 g, and 500 g)
12. Inspection mirror
13. Eccentric screwdriver

These tools are used to make the mechanical adjustments and tests outlined in Chap 8. Sony also recommends a periodic mechanical maintenance schedule of cleaning, lubrication, and replacement of parts as shown in Table 7-1.

For VHS recorders, a typical list of tools recommended by Hitachi is as follows:

1. Back tension meter
2. Cassette housing positioning jig
3. Reel disk height jig
4. Reel disk height jig adapter

5. Torque gauge
6. Torque gauge adapter
7. Servo PC board adjustment converter
8. Tension measurement reel
9. Guide roller alignment driver
10. Fan type tension gauge
11. Test/alignment tape cassette

These tools are used in making the mechanical adjustments on VHS recorders outlined in Chap. 8. The test/alignment tape is used in the electronic alignment and adjustment procedures of Chap. 9. RCA also sells a set of mechanical jigs and tools for VHS recorders as well as a test/alignment tape cassette in a package for a little over $500.

Some tools can be used on both VHS and Betamax recorders, but it is better to have a set of the special tools recommended for each format. For instance, Sony uses a torque-measuring cassette whereas the tool kit for VHS recorders includes separate torque gauges and a tension scale, The Sony torque-measuring cassette is not applicable to the servicing of VHS recorders.

Both cleaning and lubrication kits are available from Sony and Zenith for Betamax machines and from RCA and others for VHS machines. For instance, RCA markets a lubrication kit for VHS recorders for less than $20.

Test/Alignment Tapes

In their service instructions, VTR manufacturers depend heavily on test/alignment tapes, which come in cassette form and are available in the $100 range. They are particularly used for adjusting machines in the playback mode.

Betamax test/alignment tapes (available from Sony, Zenith, and others who sell Betamax recorders) have six sections in the KR5-1D version, as follows:

Section	Mode	Video Signal	Audio Signal	Playing Time
1	2H	Color bar (75%)	3kHz − 5 dB	5 min
2	2H	Monoscope	333 Hz − 5 dB	5 min
3	2H	Sweep	5 kHz − 25 dB	5 min
4	1H	Color bar (75%)	3 kHz − 5 dB	5 min
5	1H	Monoscope	7 kHz − 25 dB	5 min
6	1H	Sweep	333 Hz − 5 dB	5 min

A diagram of the 75-percent color bar signal recorded on this tape is shown in Fig. 7-2.

VHS test/alignment tapes (available from Panasonic, RCA, and others who sell VHS recorders) are set up in the following way:

Start Counter Reading:	0	039 ± 4	128 ± 6	240 ± 0
Video:	Blank	Monoscope	Color bars	Multiburst
Audio:	Blank	6 kHz	3 kHz	1 kHz

Table 7-1 Periodic Mechanical Maintenance Schedule for Betamax Recorders*

Note: C—Cleaning; L—Lubrication; R—Replacement

Name	Operating hours								
	500	1000	1500	2000	2500	3000	3500	4000	5000
VIDEO HEAD DISC assembly	C	CR	C	CRL	C	CRL	CRL	CRL	CR
VIDEO/CTL head assembly	C	C	C	C	C	CR	C	C	C
Capstan assembly		CL		CL		CL	CRL		CR
Capstan bearing assembly		CL		CL		CL		CRL	CR
Pinch roller	C	C		C		CR	C	C	C
Relay pulley		CL		CL		CRL	CL		
Drum belt		R		R		R		R	R
Capstan belt				R				R	
Relay belt				R				R	
FWD belt				R				R	
FF belt				R				R	
Counter belt				R				R	
Supply reel table assembly				CL				CL	
Take-up reel table assembly				CL				CL	
FWD idler assembly				R				R	
FF assembly				CRL				CRL	
FF idler assembly				CR				CR	
REW idler assembly				CR				CR	
Intermediate pulley assembly				C				C	
Brake band assembly				R				R	
Supply & Take-up brake assembly				R				R	
Threading brake assembly				R				R	
Ac motor					R				
Tape guides	C	C	C	C	C	C	C	C	C

*(Courtesy, Sony Corp. of America)

If one has a Sencore VA48 Video Analyzer, it is possible to make one's own test/alignment tape by following the instructions in their service manual for that instrument. By using the counter on the video recorder and switching to bar sweep, chroma bar, color bar, and crosshatch positions on the VA48, the following patterns are recorded (the recording of some audio at 1000 Hz is also suggested):

Tape Counter	Pattern
000–050	Bar Sweep
050–100	Chroma bar sweep
100–150	Color bars
150–200	Crosshatch
200–250	Bar sweep (slow speed)
250–300	Chroma bar sweep (slow speed)
350–400	Color bars (slow speed)
450–500	Crosshatch (slow speed)

Electronic Test Equipment

For field-service procedures on Betamax recorders, Sony recommends the following electronic test equipment:

1. A color monitor TV receiver
2. A dual-trace oscilloscope, with response to 10 MHz and a delay mode
3. A frequency counter (with more than four digits)
4. A VTVM
5. A VOM (20 kΩ/volt)
6. An audio signal generator
7. An audio attenuator
8. A Betamax test/alignment tape
9. An alignment tool for semifixed resistor coil

For factory procedures, Sony adds an NTSC color bar generator as a signal source.

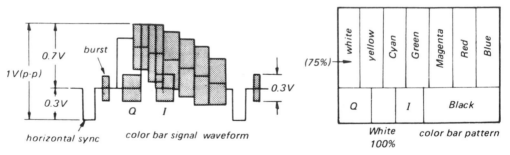

Fig. 7-2 75-percent color bar signal recorded on Betamax alignment tape (*Courtesy,* Sony Corp. of America)

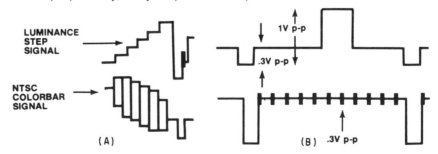

Fig. 7-3 (A) NTSC video signals, and (B) gated rainbow signals (*Courtesy,* RCA)

Fig. 7-4 Typical dual-trace scope (*Courtesy,* Hickok Electrical Instrument Co.)

For VHS recorders, RCA recommends a wider range of choice for signal generation. While they state that playback adjustments will use the test tape, either of the following two generator types can be chosen:

1. NTSC color bar generator with a standard 75-percent saturated color signal and a 5- or 10-step luminance signal (see Fig. 7-3A).
2. Gated rainbow generator capable of producing a video signal of 1 V peak-to-peak across a 75-Ω load. Required signals from this type of generator are color bars (adjusted for 0.3 V peak-to-peak chroma) and a window (superpulse) signal of 1 V peak-to-peak with 0.3 V peak-to-peak sync pulse (see Fig. 7-3B).

For the dual-trace oscilloscope and the frequency counter, RCA recommends equipment with more severe specifications, as follows:

1. A dual-trace triggered oscilloscope with LO-Cap (x10) and direct probes, a frequency response of dc ~ 20MHz, a sensitivity of 5 mV/division, and a maximum sweep rate of 0.1 μsec/division.
2. A frequency counter capable of seven-digit display and with a sensitivity of 25 mV to 5 V and a frequency range of dc to 16 MHz.

Some other VHS manufacturers, however, appear to be satisfied with dual-trace scopes having a top frequency of 10 MHz and with frequency counters of the same capability. A typical dual-trace oscilloscope and frequency counter are shown in Figs. 7-4 and 7-5, respectively.

Fig. 7-5 Typical frequency counter used in VCR alignment and adjustment (*Courtesy,* Leader Instruments Corp.)

The Use of NTSC Color Bar Signal Generators

It was mentioned above that RCA recommended the use of the test/alignment tape for playback adjustments. This appears to be a basic philosophy of most VTR manufacturers. However, they adjust the record circuits to provide matching performance when using a standard signal source. Such a source might be a generator providing 75-percent NTSC color bars with standard sync. VTR service manuals show waveforms for standard NTSC color bars. Although adjustment instructions for circuits in the record mode are given for Betamax and VHS circuitry in Chap. 9, here we will show briefly how an NTSC signal generator (such as that shown in Fig. 7-6) is used in the record-mode adjustment of luminance FM deviation, white clip adjustment, chroma circuits, and the scanner servo phase.

Adjustment of the frequency-modulated luminance signal requires that the FM modulator swing between two fixed frequency limits, which correspond to sync tip and peak white in the video signal. A common method is to set the playback video level first to the standard output level of 1 V peak-to-peak using the factory-supplied test/alignment tape cassette. The output is then terminated at 75 Ω. In the record mode, a preliminary adjustment is made to remove the action of the white-clip circuit. Then the sync-tip frequency is set in the record or E-E mode with no input video applied. A frequency counter is connected as shown in Fig. 7-7, and the sync-tip frequency is set to the proper value. A standard video signal (with 100-percent peak white) is then applied and a trial recording made. Deviation is increased in small increments while monitoring the peak-to-peak video in the modulator. A note is made of each value on one of the

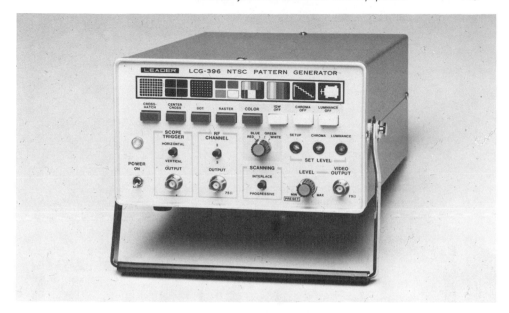

Fig. 7-6 NTSC Signal Generator (*Courtesy,* Leader Instruments Corp.)

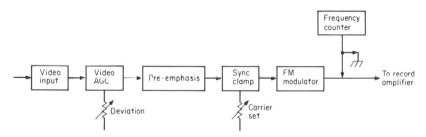

Fig. 7-7 Adjustment of FM modulator deviation levels (*Courtesy,* Leader Instruments Corp.)

audio tracks, using a microphone. The trial recording is then played back, and the input value that yields the correct 1-V, peak-to-peak, output video value is noted. Deviation is then reset to the noted value.

An alternative method is shown in Fig. 7-8. This system must be used when keyed clamps are found at the input to the FM modulator, in which case the frequency of the modulator in the absence of input video is meaningless. For this system to work, the VTR must have a true E-to-E signal path; that is, output video in the record mode must have been through the full FM modulation-demodulation process. Many home VTRs do not have full E-to-E operation. The recorder is put into the E-to-E mode with the standard color-bar signal applied. The scope is connected to monitor the output of the FM demodulator and set to observe one or two vertical fields.

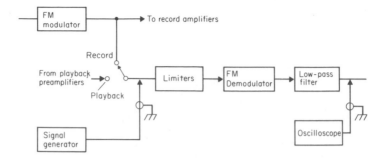

Fig. 7-8 An alternative method of adjusting FM deviation levels (*Courtesy*, Leader Instruments Corp.)

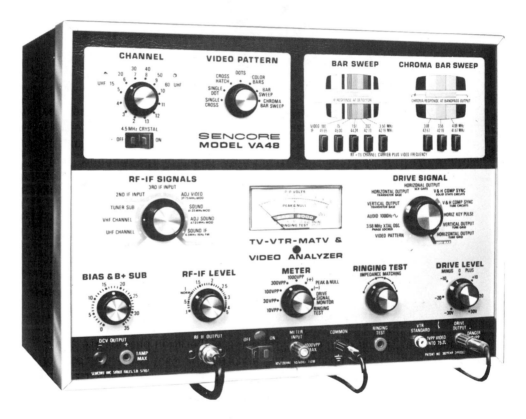

Fig. 7-9 The Model VA48 Video Analyzer offers its own unique methods for VCR maintenance (*Courtesy*, Sencore Inc.)

A CW signal from a signal generator is injected into the luminance playback circuits just ahead of the limiters and set to the sync-tip value. When the clamp level or sync-tip frequency is set correctly in the modulator, a zero beat will appear at the sync-tip level. The generator is then set to the peak white value and deviation (video amplitude) already set in the recorder to produce zero beat at the peak white level.

White Clip Adjustment

Following the deviation adjustments to the FM modulator, the white-clip adjustment must be reset to prevent the pre-emphasis spikes at the leading edges of the peak-white signal excursions from driving the modulator too high in frequency—that is, from causing overdeviation. A signal source with a 100-percent peak-white bar is needed. The crosshatch or single-cross display of an NTSC generator is the type of signal that can be used for revealing the effects of excessive FM deviation.

Chroma Circuit Adjustments

In typical helical-scan home recorders, the chroma signal is not demodulated at any point, although it is down-converted and up-converted to its original frequency. Because it is not demodulated, circuit adjustments deal primarily with absolute or relative signal amplitudes. Although burst amplitude can be used as an amplitude reference, common sources of color video signals such as TV tuners or receiver monitors are subject to variations in burst amplitude and shape because of tuner/antenna influences and the effects of multipath. For this reason, most service manuals deal with 75-percent NTSC color bars and reference chroma amplitude to the peak-to-peak value of the cyan and red bars. Where the relative value of Y and chroma must be set, as in E-to-E adjustments of luminance and chrominance values, the luminance value is set first for the standard output level of 1 V (peak-to-peak). Chroma amplitude is then set so that the tops of the yellow and cyan bars are even with the top of the 100-percent bar.

Scanner Servo Phase in the Record Mode

Servo timing is adjusted to record vertical sync at a specified point on each video track. When adjusting scanner-servo timing, the head switching point is set first, using the factory alignment tape. Sync has been recorded with great precision on these tapes. Adjustment of the servo feedback delay multivibrator is made in playback in order to put the switching point the proper number of lines ahead of vertical sync.

A similar timing adjustment is made in the record mode to make sure that vertical sync is recorded at the proper time with respect to the head-switching signal. For this adjustment, the signal source must have standard sync with correct sync and blanking intervals as well as accurate equalizing pulses and serrations. An NTSC signal generator provides such a signal. To make the adjustment, input video is monitored on one channel of a dual-trace scope. The head-switching signal—a 30-Hz square wave—is monitored with the remaining scope channel. Delayed sweep or expanded sweep is used to expand the area in the vicinity of vertical sync. Record scanner (or drum) phase is then adjusted so that the switching excursion occurs at the specified interval ahead of vertical sync.

Using a VA48 Video Analyzer

Sencore's VA48 Video Analyzer (shown in Fig. 7-9) has seven video patterns that can be selected as required for various checks. These are

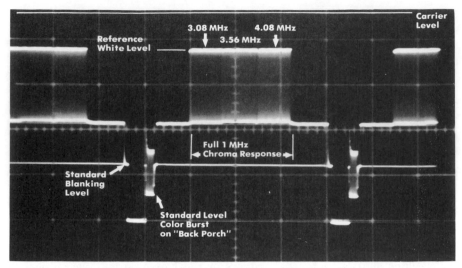

Fig. 7-10 Sencore bar sweep pattern (*Courtesy*, Sencore, Inc.)

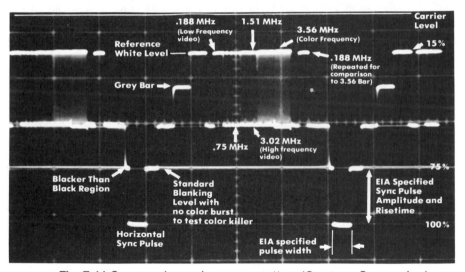

Fig. 7-11 Sencore chroma bar sweep pattern (*Courtesy,* Sencore Inc.)

single cross, single dot, crosshatch and dots, color bars, bar sweep and chroma bar sweep. As indicated previously, Sencore recommends that crosshatch, color bars, bar sweep, and chroma bar sweep be recorded on a test tape for use in video recorder servicing because these are the patterns that will be most useful for this purpose. In the color-bar position, the VA48 provides a standard 10-bar gated rainbow pattern that is similar to the output of a color-bar generator with the following improvements:

1. A color-burst signal is provided on the back porch of the horizontal sync pulse just as it would appear on a TV station signal. Most

color-bar generators do not actually provide a back-porch or separate burst signal

2. The burst signal is phase-locked to the horizontal sync pulses, thus preventing the "cog-wheel" effect caused by most color-bar generators and resulting in a more stable pattern for color alignment and troubleshooting.

Bar Sweep

The patented bar-sweep pattern, shown in Fig. 7-10, is used for all luminance circuit testing. Since this pattern does not have a color-burst signal, it gives a positive test for all automatic color killer circuits found in both the record and playback sections. These circuits should automatically switch to the black-and-white mode when the color burst is not present. The bar-sweep pattern produces various amounts of luminance information. The gray scale at the left of the switchable frequency bars, for example, checks the black, gray, and white signal levels for proper amplitudes, linearity, and clipping. Each frequency bar after this gray scale alternates between pure black and pure white levels to test the circuits for proper frequency response. The bar-sweep pattern looks much like the familiar multiburst pattern used in many types of video tests. One difference, however, is that the bar sweep is made up of square waves, whereas the multiburst is composed of sine waves. The key difference is that the bar sweep shows circuits that will ring on fast signal transitions, whereas the multiburst signal may be passed without any noticeable ringing. Because of this characteristic, the bar-sweep pattern can check the video frequency response of the entire record/playback system.

The Chroma Bar Pattern

The chroma bar pattern contains a standard-level color burst for operation of the color killer circuits and the automatic color control (ACC) circuits. Its waveform is shown in Fig. 7-11. The color information is phase-locked back to the horizontal sync pulses so that the pattern produces a line-by-line phase inversion just like that of an off-the-air signal. This is a very important feature because it makes it possible for the comb filters used in almost all playback circuits to detect the proper color signal as opposed to the unwanted crosstalk information from an adjacent video stripe on magnetic tape. Since the chroma bar sweep covers the entire range of color information—that is, ±500 kHz on either side of the color subcarrier—this pattern is ideal for testing all chroma conversion circuits to make sure that they are not restricting some of the color detail information.

Sencore provides a manual that describes the step-by-step electrical alignment of the various circuits in both Betamax and VHS recorders, much as the VTR manufacturers describe them in their own service manuals. Whereas the latter tend to be more general in their reference to signals, Sencore is quite specific about their VA48.

8

Care and Maintenance of the Mechanism

Before the recorder is disassembled for trying mechanical adjustments, it is a good idea to look for the cause of any symptoms indicated. Simple operational troubles may be due to equally simple causes, such as those listed in Table 8-1. Beyond these simple failures of equipment and of the user to operate certain controls, failure to provide adequate cleaning of video heads and other parts can easily be the reason for lack of performance. Table 8-2 provides a guide for cleaning that will insure optimum performance if followed on a regular schedule.

Table 8-3 provides a troubleshooting guide that pin-points those electrical components that most frequently cause operational failures. For instance, a failure of the machine to load tape in the playback mode may be due to a micro-switch. From all appearances one might think the cause to be a mechanical one, but most of the mechanism is electrically or electronically controlled or operated. Faulty controls may make it impossible for the mechanism to function.

The service manuals of recorder manufacturers devote at least one chapter to adjustments and replacements in the mechanical portion of the VCR. One should follow their specific instructions whenever possible. However, some of the more general procedures for VHS and Betamax units (for example, measurement, and adjustment of tape tension, replacement of video heads, tape path adjustment, and reel table height adjustment) may be covered in such a way that one can grasp the basic steps involved for all units. The service manual of any recorder in question will provide the specifics required for that job.

Video Head Cleaning

Video heads may be cleaned by two different methods: (1) with cleaning fluid and a chamois-tipped cleaning stick, or (2) with a cleaning cassette. The cleaning fluid for the first method is usually obtained from cleaning kits supplied by the manufacturer of the video recorder or others who sell such kits separately. Isopropyl alcohol is a typical cleaning fluid for this purpose. Methanol is a more effective cleaner but can be a health

Table 8-1   Guide for Operational Troubles

Symptom	Cause	Action to Take
No operation at all	(A) Power switch is not pressed. (B) Timer switch is set to On position.	(A) Press power switch to turn the recorder On. (B) Set timer switch to Off position.
Operation buttons do not lock although they can be pressed in.	(A) Dew indication lamp is illuminating. (B) Power indication lamp is not illuminating.	(A) Wait until the lamp goes off. (B) Check the ac power cord.
Auto stop during rewind	(A) Tape counter memory switch is set to On position. (B) Tape is damaged.	(A) Set this switch to Off position. (B) Cut out damaged portion of tape and splice with splicing tape.
Record button is locked and cannot be pressed.	Tab on the cassette is broken.	Cover hole with cellophane tape.
On-the-air TV programs cannot be watched	TV/VTR selector on panel is set to the VTR position	Set the selector to the TV position.
On-the-air TV programs cannot be recorded	(A) Input selector on the panel is set to the camera position. (B) Antenna cable is poorly connected.	(A) Set the selector to the TV position. (B) Connect the cable properly.
No sound during an on-the-air TV program recording	Microphone is connected to the mike jack on the front panel.	Disconnect the microphone.
No playback image	TV/VTR selector is set to TV position.	Set the selector to the VTR position.
Noisy picture during playback	Tracking control on panel is misadjusted.	Adjust the control correctly.

Table 8-2   Guide for Head Cleaning

Symptom	Cause	Action to Take
Snow noise on a playback image	Video head clogging	Clean video heads (See Fig. 8-1) with a cleaning stick as shown in Fig. 8-2.
Periodic noise on a playback image	Control head clogging	Clean control head with a cleaning stick as shown in Fig. 8-3(D).
Little or no sound	Audio head clogging	Clean audio head with a cleaning stick as shown in Fig. 8-3(D).
Unstable tape transportation	Parts need cleaning	Clean the parts indicated in Figs. 8-4, 8-5, and 8-6.

Table 8-3   Troubleshooting Guide for the Mechanism

General Symptom	Detailed Symptom	Action to Take
No operation of playback, rewind, and fast forward	(A) Power indication lamp doesn't illuminate.	(A) Check the power supply and regulator circuits. Check the transformer.
	(B) Possible to lock the fast forward or rewind button, but no tape movement.	(B) Check the capstan motor.
	(C) When play button is pressed, the loading operation starts, but within a few seconds it stops.	(C) Check the capstan motor.
	(D) When the power switch is turned On, the pressure-roller solenoid works, but within a few seconds it is released and there is no operation.	(D) Check the loading completion switch.
	(E) Each button is locked, but there is no operation.	(E) Check the reel belt.
	(F) Each button is not locked, but power indication lamp is illuminating.	(F) Check the sensor lamp.
	(G) Dew indication lamp is illuminating.	(G) Check the dew sensor.
No rewind operation	(A) Rewind button is not locked.	(A) Check the supply photo transistor.
	(B) When a tape reaches its end (tranlucent portion), there is no automatic stop.	(B) Check the supply photo transistor.
	(C) Rewinding speed is very slow.	(C) Check reel belt, rewind roller, and tape itself.
	(D) Rewind button can be locked, but there is no tape movement (although play and fast forward modes are normal).	(D) Check rewind switch on circuit board.
	(E) Rewind button can be locked, but there is no operation of any kind (including play and fast forward modes).	(E) Check the reel belt.

Table 8-3   Troubleshooting Guide for the Mechanism (cont'd)

General Symptom	Detailed Sympton	Action to Take
No fast forward operation	(A) Fast forward button is not locked.	(A) Check the take-up photo transistor.
	(B) When a tape is rewound completely to its translucent portion, there is no automatic stop.	(B) Check the take-up photo transistor.
	(C) Fast forward speed is very slow.	(C) Check the reel belt, rewind roller, fast forward idler, and the tape itself.
	(D) Fast forward button can be locked, but there is no operation (although rewind and play modes are operating correctly).	(D) Check the fast forward switch.
	(E) Fast forward button can be locked, but there is no operation of any kind (including rewind and play modes).	(E) Check the reel belt.
No playback operation	(A) After loading completion, the unit automatically stops within a few seconds, and the tape comes back into the cassette case. Picture and sound are normal for a few seconds before the above trouble.	(A) Check the reel sensor, counter belt, play roller, and reel belt. Check the pressure roller solenoid and capstan belt.
	(B) After loading completion, the unit stops automatically within a few seconds, and there is no unloading operation.	(B) Check the unloading completion switch.
	(C) After loading completion, loading motor is still rotating. At that time there is no sound and image.	(C) Check the loading completion switch
	(D) No loading operation.	(D) Check the play switch and the tape-slack microswitch.
	(E) After loading completion, the tape movement stops, although the capstan motor is rotating.	(E) Check the pause switch.

Table 8-3   Troubleshooting Guide for the Mechanism (cont'd)

General Symptom	Detailed Sympton	Action to Take
Abnormal loading or un-loading operation	(A) No loading or unload-ing.	(A) Check the loading belt and the loading motor.
	(B) During playback, no unloading operation occurs when the stop button is pressed, and the power indication lamp turns off after 6 or 7 sec.	(B) Check the play switch.
	(C) When the power switch is turned On, the load-ing motor rotates and doesn't stop at all.	(C) Check the unloading completion switch.
No automatic stop opera-tion	(A) There is no automatic stop when a tape is forwarded to its end (translucent portion) on play or fast forward modes.	(A) Check the take-up photo transistor.
	(B) When a tape is re-wound completely (to translucent portion), there is no automatic stop.	(B) Check the supply photo transistor.
No tape counter operation	(A) Tape counter is not ro-tated.	(A) Check the counter belt.

hazard should it happen to get on the skin of the user. For the first method, the following procedure is used:

1. Press the Eject button and remove the tape cassette.
2. Remove screws from the top panel and lift the panel off.
3. Wet the chamois-tipped cleaning stick with cleaning fluid.
4. Turn the motor by hand in the direction of the arrow shown in Fig. 8-1. Never try to clean the video heads with the motor running, and never wipe them vertically. Always wipe them from side to side.
5. Touch the head area lightly with the chamois tip while turning the drum about three times (see Fig. 8-2).

There is a real danger of damaging video heads if pressure is applied. Since they are only about 40 microns (0.04 mm) in size, they must be handled with care.

Cleaning Cassettes

The video-head cleaning method described above has been recom-mended by home VCR manufacturers, some of whom have issued warranty

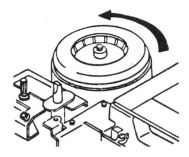

Fig. 8-1 Manual rotation of video head drum to prepare for head cleaning (*Courtesy,* Sony Corp. of America)

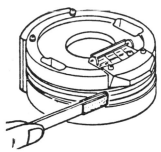

Fig. 8-2 Manual cleaning of video head (*Courtesy,* Sony Corp. of America)

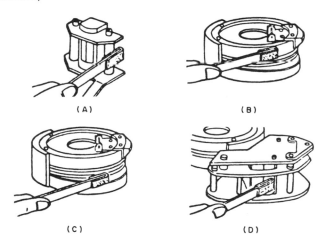

Fig. 8-3 Manual cleaning of some other parts in the tape path: (A) Entrance guide (erase head), (B) upper drum (surface contacting the tape), (C) lower drum (surface contacting the tape), and (D) CTL, audio erase head (head surface) (*Courtesy,* Sony Corp. of America)

restrictions against the use of other cleaning methods. Cleaning cassettes have become increasingly popular, however, and some of them have been endorsed by recorder manufacturers. Made to match in size and fit in place of standard tape cassettes, they allow video heads to be cleaned without entering the machine with cleaning sticks or swabs.

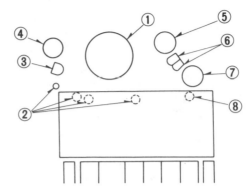

Fig. 8-4 Top view of VHS parts to be cleaned when tape transportation is unstable: (1) Cylinder, (2) posts, (3) full erase head, (4) supply inertia roller, (5) takeup inertia roller, (6) audio control and audio erase heads, (7) pressure roller, and (8) capstan shaft (*Courtesy,* Sylvania)

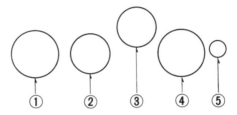

Fig. 8-5   VHS parts under cassette compartment to be cleaned when tape transportation is unstable: (1)Supply reel, (2) rewind roller, (3) fast forward idler, (4) takeup reel, and (5) play roller (*Courtesy,* Sylvania)

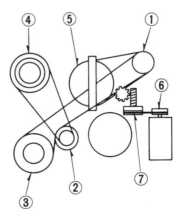

Fig. 8-6 Bottom view of VHS parts to be cleaned when tape transportation is unstable: (1) capstan motor pulley, (2) intermediate pulley, (3) main pulley, (4) play pulley, (5) capstan flywheel, (6) loading motor pulley, and (7) pinion gear pulley (*Courtesy,* Sylvania)

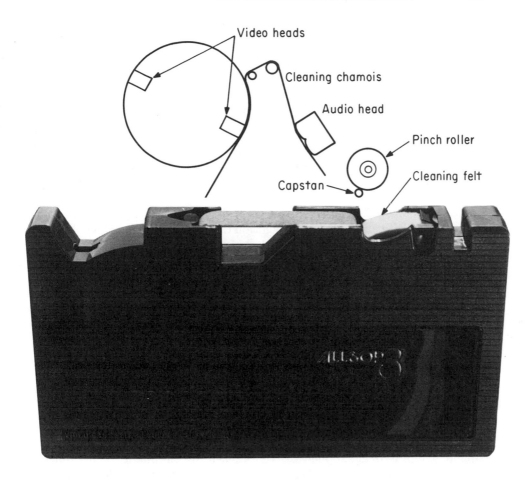

Fig. 8-7 Cleaning cassette (*Courtesy,* Allsop)

Such cassettes perform their function with either an abrasive or a solvent method. In the abrasive method, debris on the heads is lightly scraped away; in the solvent method, the debris is dissolved and carried away as the cleaning tape passes over the video head. A typical example of an abrasive cleaning cassette is the one offered by the manufacturers of Scotch tape. Solvent (or wet type) cassette cleaners are offered by TDK, Fuji, and Allsop.

Although abrasive cassettes have not been accepted as readily as solvent ones by recorder manufacturers, the Scotch unit is unique in that it delivers a flashing message on the video screen: "When you read this message, the heads are clean. Stop the recorder now." If this message does not appear within 30 sec, it is to be assumed that the Scotch cassette has not been able to complete the cleaning job and the recorder should be stopped and its cleaning problem examined. Obviously, video heads should not be exposed to an abrasive cleaning tape for any extended period of time. Scotch indicates, however, that a 5-min pass of their cleaning tape has the equivalent effect of playing a regular video recording tape for a period of 1 hr insofar as head wear is concerned.

The Fuji solvent-type cleaning tape requires a 10-sec pass at any of the standard video tape speeds to complete a cleaning cycle. A second 10-sec pass is allowed if necessary to complete a more difficult job. However, no more than three such passes are to be made so as to limit exposure of the video heads to the solvents.

The Allsop solvent cleaning cassette cleans the audio head, the capstan, and the pinch roller as well as the video heads. It includes two cleaning fabrics, one of which is a nonabrasive chamois and the other felt. The chamois cleans the video and audio heads, and the felt cleans the capstan and pinch roller, following the path shown in Fig. 8-7. These fabrics are saturated with an alcohol freon solution that is available in refill bottles. The Allsop cassette shuts off after 4 or 5 seconds of cleaning action to protect the tape deck from excess exposure to chemicals.

Measurement and Adjustment of Tape Tension

It is desirable to be able to record programs on one machine and play them back on another, but it is essential that tape tension be maintained at the proper level in both recorders if this degree of compatibility is to be realized. If the recorder on which the playback is being made has more or less tape tension than the recorder on which the recording was made, the vertical lines of the picture will be skewed to the left or right, respectively, depending upon the tension variation.

Tape tension is measured by at least three types of gauges. A very popular gauge is that supplied by Tentel, bearing the name, Tentelometer*. It is, in effect, a contact-type gauge in that it measures in-line tension when its probes are slid over the tape. Other types of tension gauges mentioned in VTR service manuals include the sector type and the cassette type. Various types of tension are measured as described below.

Measurements with Tentelometers

Since this type of gauge measures tension by detecting the force against its center probe, it cannot differentiate between tape stiffness and tape tension. Tape stiffness must therefore be excluded from the reading if pure tension is to be measured. To do so, it is necessary to suspend a known weight (furnished with the Tentelometer) from a short length of tape that is similar to that whose tension is to be measured (see Fig. 8-8). The Tentelometer is set to read the value of the weight, thus in effect taking account of the tape stiffness. With the weight suspended on the sample of tape, the gauge should be slid over the tape and moved up and down slightly to note the average reading of the pointer. Maximum accuracy requires that the calibration be made with the tape moving in the same direction through the probes as it will be moving in the recorder.

For the measurement of tape tensions on Betamax recorders the following steps should be taken:

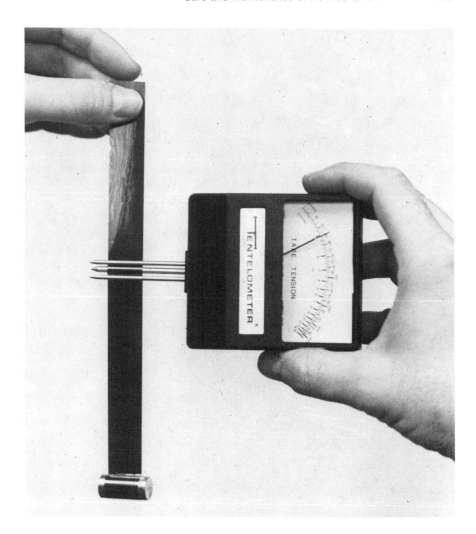

Fig. 8-8 Calibration of Tentelometer to eliminate tape stiffness factor (*Courtesy,* Tentel Corp.)

1. Calibrate the Tentelometer on a sample of ½-in. Beta tape similar to the tape being used in the machine whose tape is under test (this procedure is described above.)
2. At a minimum, supply (hold-back) and take-up tension should be checked. Using a 2-hr Beta tape in a full supply reel, measure the supply tension with the machine in either the play or record mode, as shown in Fig. 8-9. The supply tension should measure 15 to 20 gm, and the take-up tension should measure 55 to 110 gm.
3. In addition, the Tentelometer can be used to make torque measurements, such as the rewind torque and the fast forward torque. The stalled rewind torque can be measured by stalling the tape where it comes out of the cassette and measuring the tension that the machine exerts on the tape. This tension should be 180 to

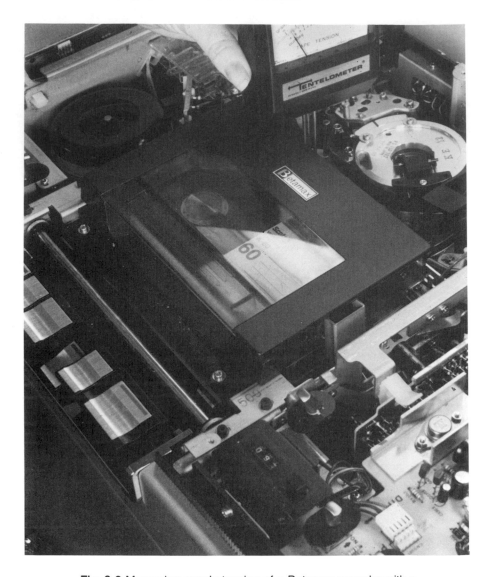

Fig. 8-9 Measuring supply tension of a Betamax recorder with a Tentelometer (*Courtesy,* Tentel Corp.)

200 gm. The stalled fast forward torque can be measured by shuttling the tape onto the take-up reel (until almost full) and by stalling the tape where it comes out of the cassette. Then the force (tension) exerted by the take-up reel is measured at point T/U in Fig. 8-10. This tension should be 220 to 260 gm. The nonstalled rewind and fast forward tension should be 5 to 15 gm going into the cassette to assure proper tape packing without cinching, binding, or edge damage.

For the measurement of tape tensions on VHS recorders, the following steps should be taken:

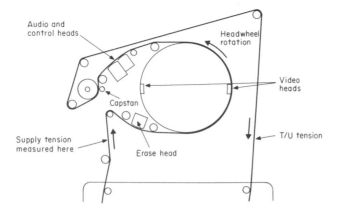

Fig. 8-10 Points in tape path where supply and take-up tension is measured in Betamax recorder using Tentelometer (*Courtesy,* Tentel Corp.)

Fig. 8-11 Measuring supply tension of a VHS recorder with a Tentelometer (*Courtesy,* Tentel Corp.)

1. Calibrate the Tentelometer on a sample of ½-in. VHS tape similar to the tape being used in the machine whose tape is under test. This procedure is described above and illustrated in Fig. 8-11.
2. As a minimum check of the supply (hold-back) tension, take the following steps:
   (a) Remove the rear panel and top panel of the VHS recorder.

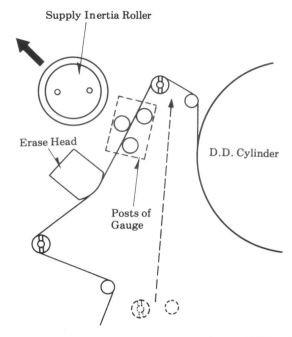

Fig. 8-12 Procedure for measuring supply tension of VHS unit with a Tentelometer (*Courtesy,* Sylvania)

(b) Pull the supply inertia roller in the direction indicated by the arrow in Fig. 8-12, and hold it back by adhesive tape.

(c) Play back the cassette tape, and wait until the tape running has well stabilized.

(d) Set the posts of the tension meter in front of the supply inertia roller, as indicated in Fig. 8-12, and take the meter reading, which should be between 20 and 25 gm to insure interchangeability between recorders. To adjust the tension, move the threading spring hook (shown in Fig. 8-13) as follows: (1) Loosen screw A and set a fine-adjustment screwdriver in hole B. (2) Move the threading spring hook in either direction indicated by the arrow to vary the tension. (3) Upon completion of adjustment, tighten screw A and remove the adhesive tape from the supply inertia roller.

3. The remaining tension and torque measurements on VHS recorders require the use of a modified cassette (Fig. 8-14). Such a cassette can be made from a 1-hr (2-hr LP) VHS cassette by removing the front door of the cassette, prying off the top half of the take-up reel, and drilling and filing a hole in the cassette top cover to allow the gauge to be inserted over the take-up side of the tape (after the pinch roller and capstan). It will be necessary to remove the cassette chamber top cover of the recorder to allow access to the modified cassette during tension measurements. In addition, a shim ⅛ in. thick by 7⅝ in. long is required on the

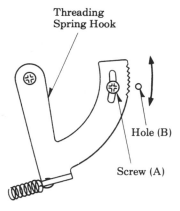

Threading
Spring Hook

Hole (B)

Screw (A)

Fig. 8-13 Procedure for adjusting supply tension of VHS machine (*Courtesy,* Sylvania)

Fig. 8-14 Modified cassette used with Tentelometer to measure various VHS tensions (*Courtesy,* Tentel Corp.)

Panasonic, RCA type recorders to hold the cassette in the proper position when the cassette chamber top cover is removed.

4. Using the modified cassette (Fig. 8-15), take-up tension should be 26 to 52 gm, with a 1-hr tape (2-hr LP). Stalled rewind tension should measure 65 to 80 gm, and stalled fast forward tension should be 100 to 125 gm. Here again, the tape is stalled by using finger pressure over a clean, lint-free cloth where the tape comes out of the cassette, and the Tentelometer is inserted over the tape. This measurement replaces the torque gauge readings, which may be made by other means in the VTR service manuals. Unstalled rewind and fast forward tensions can be measured directly with the modified cassette. They should range from 3 to 10 gm.

Fig. 8-15 Using modified cassette in VHS recorder for tension measurements (*Courtesy,* Tentel Corp.)

The Use of a Sector Tension Gauge

Figure 8-16 shows the use of a sector type gauge to measure back tension in a Betamax recorder. A forward (FWD) back tension measurement fixture (3) is placed on the supply-reel table assembly (4). The tape is threaded as shown and withdrawn over the chassis (5) at the right side of the machine. There the tape is pulled with the sector gauge at a speed of 4 cm/sec. A 40- to 50-gm reading should be obtained on the gauge.

If the tension reading is not within these limits, loosen the adjusting screw (1), and move the tension regulator adjusting spring hook (2) in the direction indicated by the arrow. If the tension regulator spring hook is

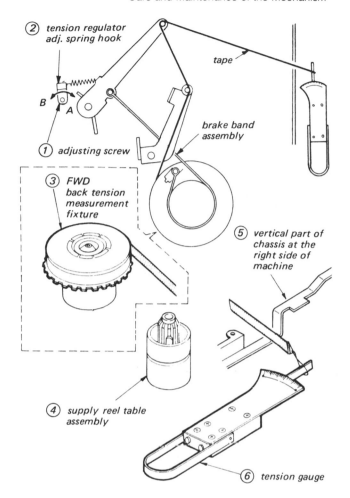

Fig. 8-16 Procedures for Betamax back tension measurements with sector gauge (*Courtesy,* Sony Corp. of America)

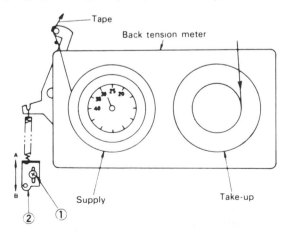

Fig. 8-17 Procedure for use of cassette-type tension gauge (*Courtesy,* Hitachi Sales Corp. of America)

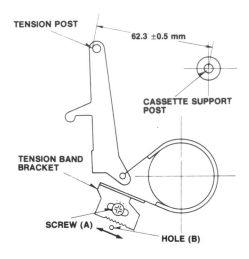

Fig. 8-18 Back tension band adjustment (*Courtesy,* Sylvania)

moved in the direction of A, the back tension is reduced. If moved in the direction of B, the back tension is increased.

The manufacturer's service manual gives complete details of disassembly of the recorder to make possible these measurements and adjustments.

The Use of a Cassette Tension Gauge or Meter

Figure 8-17 shows a cassette type tension gauge, which replaces the regular tape cassette. For a VHS machine, the manufacturer recommends that the meter should read from 22 to 32 gm-cm in the playback mode. The service manual for this VHS recorder states that the following adjustments should be made:

1. If the tension is greater than 27 gm-cm, loosen screw (1) and move spring hanger (2) in the direction of arrow A and tighten (1) when the meter reads 27 gm-cm.
2. If the tension is less than 27 gm-cm, loosen screw (1) and move spring hanger (2) in the direction of B to obtain 27 gm-cm.

Note: Except in the case of the Tentelometers, it should be noted that various tension readings have been obtained above. These readings depend upon the type of tension gauge and the method of measurement, with variations between the different designs of recorders. It is essential, therefore, that one refer to the service manuals of the specific recorder being tested for the exact steps and readings.

In both Betamax and VHS recorders, back tension is controlled by brake band assemblies of the type shown in the upper part of Fig. 8-16 and separately in Fig. 8-18. The adjustment for back tension has already been

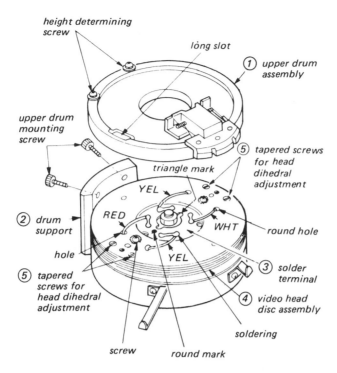

Fig. 8-19 Betamax video head disk assembly (*Courtesy,* Sony Corp. of America)

covered. However, the adjustment of the back tension band in a VHS recorder is as follows:

1. Remove the cassette holder assembly.
2. Short anode of regulator diode to ground.
3. Place instrument in play mode with no tape in the machine. After loading operation is completed, turn ac power Off and press the play button again.
4. Loosen screw A as shown in Fig. 8-18.
5. Insert tension adjust screwdriver in hole B, and carefully adjust the distance from the tension post to the center of the cassette support post. It should be as shown in the figure, or 62.3 ± 0.5 mm. Tighten the screw after the adjustment has been made.

Video Head Replacement

Because of wear caused by their contact with the tape, video heads are likely to require replacement. Excessive wear will result in a drop in their output level. Other possibilities leading to replacement include broken connections to the rotary transformer and dirty switch contacts. The latter can often be repaired without removing the heads themselves. Since video heads are costly to replace, one should be very sure that it is absolutely necessary before doing so.

In both Betamax and VHS recorders, the video heads are replaced as a disc assembly containing both of them. In the Betamax, an upper drum assembly must be removed before the video head disc assembly can be replaced. In VHS machines, however, the video head disc assembly can be removed directly. In VHS machines, this assembly is often called the upper DD (for Direct Drive) cylinder.

Figure 8-19 illustrates the steps in replacing the video heads of a Betamax recorder, described below:

1. Remove the upper drum mounting screws with an Allen wrench, and remove the upper drum assembly (1) from the drum support.
2. Remove the four wires of the video head disc from solder terminal (3).
3. Remove the screws and the video disc assembly (4).
4. Clean the bottom and flange surfaces of the replacement video head disc assembly with a piece of alcohol-dampened cloth.
5. Place the replacement video head disc assembly (4) so that the red lead of the disc assembly will be close to the small round mark on the solder terminal (3), and secure the disc assembly temporarily with two screws. This temporary installation is made so that an

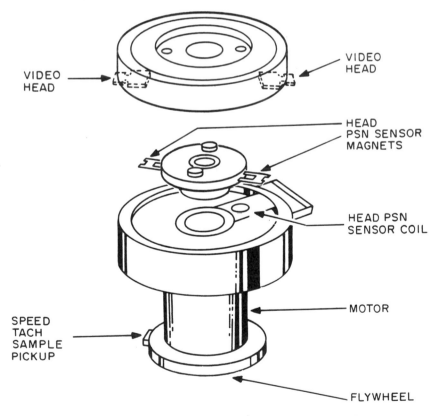

VIDEO HEAD

VIDEO HEAD

HEAD PSN SENSOR MAGNETS

HEAD PSN SENSOR COIL

MOTOR

SPEED TACH SAMPLE PICKUP

FLYWHEEL

Fig. 8-20 VHS video head disk and D-D cylinder assembly (*Courtesy, RCA*)

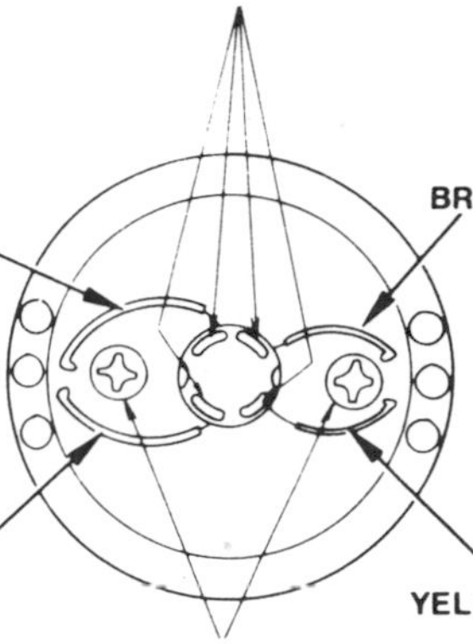

LEADS FROM ROTARY TRANSFORMER

YELLOW　　　　　　　　BROWN

RED　　　　　　　　　YELLOW

UPPER DD CYLINDER (VIDEO HEADS)

Fig. 8-21 Connection to VHS video heads (*Courtesy*, RCA)

eccentricity check and adjustment can be made without additional disassembly work.

6. Solder the leads of the video head to solder terminal (3). Bend the video head leads flat on the top surface of the video head disc assembly (4) so that the slack of the leads does not touch the upper drum assembly (1).

7. Set the upper drum assembly (1) to drum support (2). Tighten the mounting screws with an Allen wrench while holding the height-determining screws with the fingers.

8. To optimize the head disc surface, run the lapping cassette tape in the play mode for about 5 sec.

Figures 8-20 and 8-21 illustrate the replacement of the video heads in a VHS recorder. Figure 8-20 shows that there is no upper drum assembly above the part containing the video heads. As mentioned previously, the part containing the video heads in VHS recorders is referred to as the upper DD cylinder, the connections of which are shown in Fig. 8-20. The steps taken in removing and replacing the upper DD cylinder are as follows:

1. Unsolder the four leads, which are color-coded to match the leads from the rotary transformer that is part of the DD cylinder.

2. After removing a screw from the static discharge brush assembly and two retaining screws from the top of the upper cylinder, very gently lift the upper cylinder from the main DD cylinder shaft. Do not touch the video heads.

3. Before replacing or reinstalling the upper cylinder, clean the DD cylinder shaft and the inside of the upper cylinder with a "Kimwipe" and solvent. The upper cylinder should fit snugly on the DD motor shaft. If the shaft and the mating surface of the upper cylinder are clean and properly aligned, the upper cylinder will slide onto the shaft without undue pressure being exerted.

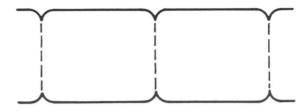

Fig. 8-22 VHS balanced head output (*Courtesy,* RCA)

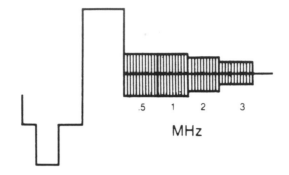

Fig. 8-23 Output of multiburst signal in VHS head equalization (*Courtesy,* RCA)

4. After replacing the upper cylinder, use "Kimwipes" and solvent to clean the surface of the upper cylinder, exercising care not to touch the video heads.

When video heads are replaced, it is necessary to confirm their equalization and PG shifter adjustments. Head balance for VHS recorders may be easily checked by using the following procedures:

1. Connect channel-1 scope probe (0.5 V/div.; 5 ms/div.) to head amplifier output.
2. Record a color program and play it back.
3. Adjust balance control for equal head output (output level like that shown in Fig. 8-22 should be obtained).

Video Head Equalization Confirmation

These adjustments affect the frequency response of the two video head playback preamplifiers. When the upper cylinder (video head assembly) is replaced or the preamplifier circuitry is serviced, the overall "head system" frequency response and matching should be checked by playing back the multiburst signal on the test tape. The signal bursts viewed at the video output (0.5 V/div.; 10 s/div.) should be within the limits shown in Fig. 8-23, and there should be minimum flicker of the 2- and 3-MHz bursts.

If the conditions of Fig. 8-23 are met, no further checks or adjustments of the luminance and color circuits are required after video head replacement. Improper response or poor picture quality indicate the need for head equalization. If this adjustment is required, one should refer to the appropriate service manual because test points and component designations vary depending on the model to be serviced.

The P.G. (Pulse Generator) Shifter Adjustment

If further checking following video head replacement is indicated, video head switching points should be examined. The P.G. shifter determines the video head switching point during playback. Misadjustment of either pulse generator control can cause head-switching noise in the picture and/or horizontal jitter.

VHS service manuals usually do not include instructions for eccentricity or dihedral error adjustments. It is assumed that VHS manufacturers prefer to do these adjustments in the factory rather than in the field. Betamax service manuals, however, do include procedures for eccentricity and dihedral adjustments that are to be made after video head replacement.

Betamax Video Head Disk Eccentricity Adjustment

For the eccentricity adjustment, proceed as follows:

1. Remove threading arm guide (I) in Fig. 8-24.
2. Install the eccentricity gauge (2) with thumbscrew (II).
3. Set the eccentricity gauge with thumbscrew (I) so that its probe contacts the head disk circumference about 2 mm below the top edge of the video head disk assembly [see (3) in Fig. 8-25].
4. Rotate the ac motor slowly counterclockwise. Adjust the video head disk assembly eccentricity—by very gently tapping its circumference with the handle of a screwdriver—so that the gauge reading deflection is within 3 microns (see Fig. 8-25).
5. When the eccentricity is less than 3 microns, finger tighten the two mounting screws alternately and finally tighten them fully with a screw driver. The tightening torque must be more than 10 kg/cm.
6. Make a final test of eccentricity using the gauge after the screws are fully tightened.

After the adjustment, solder the leads of the video head to solder terminal (3) (see Fig. 8-19). Bend the video head leads flat on the top surface of the video head disk assembly (4) so that the slack of the leads does not touch the upper drum assembly (1). Set the upper drum assembly to drum support (2). Tighten the mounting screws with an Allen wrench while holding the height-determining screws with the fingers.

To optimize the head disk surface, run the lapping cassette tape in the play mode for about 5 sec.

Betamax Video Head Dihedral Adjustment

This adjustment may not be necessary, of course, if no dihedral error is present. If there is dihedral error, however, proceed as stated (referring to Fig. 8-19):

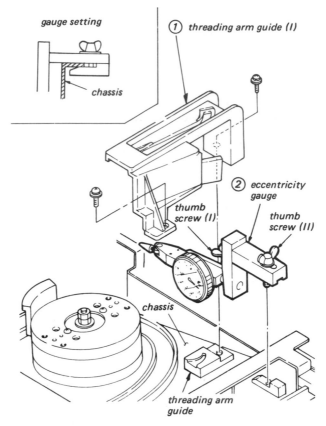

Fig. 8-24 Betamax drum eccentricity measurement setting (*Courtesy,* Sony Corp. of America)

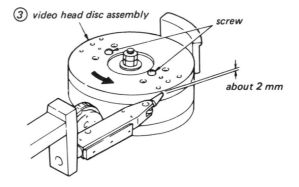

Fig. 8-25 Betamax drum eccentricity measurement (*Courtesy,* Sony Corp. of America)

1. Play back the monoscope signal segment of the alignment tape, and check for dihedral error at the top of the picture (or at the bottom, at the switching point.)
2. If dihedral error is observed, install four tapered screws (5) in

each of the four holes on the video head disc assembly (4) through the slots in the upper drum assembly (1).

3. Turn one of the four screws until the screw touches the head base. Give this screw one additional quarter turn.
4. Play back the monoscope signal of the alignment tape, and check again for dihedral distortion.
5. If the symptom is worse, unscrew the screw identified in step (3) above, and turn down the screw on the opposite side of the same video head. Repeat, in quarter turn steps, to eliminate the dihedral error.
6. Perform electrical adjustment.

Interchangeability Confirmation

Interchangeability in video recorders refers to the ability of a recorder to play back tapes recorded on another machine. This capability may be affected by the replacement of video heads, the DD motor, tape guides, control track/audio heads, and so forth. Some VCR manufacturers do not recommend any procedures for interchangeability checks or adjustments, preferring that defective units be returned to their factory for adjustment or repair. Others, however, describe the procedures to be taken if interchangeability adjustments must be made in the field. RCA outlines an "Interchangeability Confirmation" procedure for VHS recorders. They state that if this test can be passed, there will be no need for interchangeability adjustments.

The RCA Interchangeability Confirmation proceeds as follows:

1. Connect scope channel 1 (5 V/div.; 5 ms/div.) to the test point providing the tracking shifter voltage-trigger on channel-1 signal.
2. Connect scope channel-2 (20 mV/div.) to the test point providing the head amplifier output.
3. Play the monoscope signal on the test tape and confirm that the tracking fix pulse is 19 ms when the tracking control is at the fixed position. Adjust the tracking fix resistor for proper pulse width if necessary.
4. Adjust the tracking control for maximum RF signal amplitude from the head amplifier output at the center of the envelope.
5. Adjust the scope vertical gain control so that maximum envelope amplitude is four graticule divisions.
6. Turn tracking control clockwise until any part of the envelope drops to 50 percent—two graticule divisions.
7. Adjust the scope horizontal position control to place moveable edge of tracking fix pulse (at the tracking shifter test point) on graticule line, and note its position.
8. Turn tracking control counterclockwise until any part of the envelope amplitude drops 50 percent—two graticule divisions.
9. Determine the tracking fix pulse width change between the two 50-percent points. This change should be greater than 7 ms; if it is, tape guide adjustment is not required. If the change in pulse width is less than 7 ms, tape guide adjustment is required.

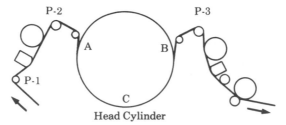

Fig. 8-26 Tape passing around three adjustable posts (*Courtesy, Sylvania*)

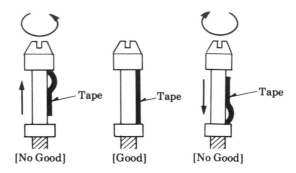

Fig. 8-27 Effects of tape post positions on VHS tape path (*Courtesy, Sylvania*)

10. Set the tracking control to fixed position. Then, adjust the control track/audio head assembly position to obtain the maximum RF envelope at the head amplifier output in the detent fixed position.

Tape Interchangeability Pre-Adjustment for Posts and A/C Head Height (VHS)

Figure 8-26 shows the three posts around which the tape travels (use a normal cassette tape for pre-adjustment to avoid damage to alignment tape). If these posts are incorrectly adjusted, the tape will take the appearance shown at the left and right in Fig. 8-27. Each post should be adjusted by a special post adjustment screwdriver (available from the VHS recorder company). The use of this screwdriver is illustrated in Fig. 8-28. If curling of the tape is apparent, the posts should be adjusted to eliminate it (as indicated above).

As a further step in this pre-adjustment procedure, adjust the two screws (A) and the screw (B) in Fig. 8-29 so that the audio/control (A/C) head height relative to the tape is such that the tape passes by this head as shown in Fig. 8-30.

More Precise Tape Interchangeability Adjustment (VHS)

1. Connect an oscilloscope to the head amplifier output test point.
2. Play back the monoscope portion of the alignment tape.

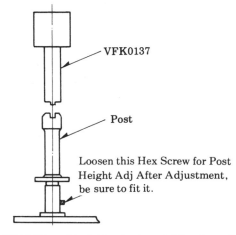

Fig. 8-28 VHS post height adjustment (*Courtesy,* Sylvania)

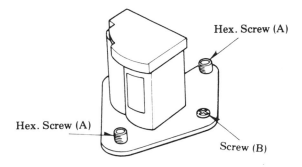

Fig. 8-29 VHS audio/control head showing height adjustment screws (*Courtesy,* Sylvania)

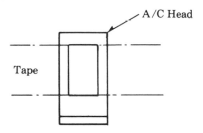

Fig. 8-30 Correct tape path past VHS audio/control head (*Courtesy,* Sylvania)

Tape Crease and Slack Adjustment

1. Referring to Fig. 8-25, watch the portions of the tape between Post P-2 and A and between Post P-3 and B carefully.
2. If the tape creases or slack appears at these portions, adjust the position of the inclined base as shown in Fig. 8-31. To adjust it,

loosen screw (A) slightly, set an eccentric screwdriver to hole (B), and move the base as shown in Fig. 8-31.

Post Height Adjustment

While playing back the monoscope portion of the alignment tape, adjust posts P-1, P-2, and P-3 while watching the scope display at the head amplifier output test point. They should be adjusted so that the RF envelope on the scope becomes as flat as possible. Referring to Fig. 8-32, $V1/V \geqslant 0.7$ and $V2/V \geqslant 0.7$. Proceed as follows:

1. When the scope display appears as in Fig. 8-33, adjust the height of P-2.
2. When the scope display appears as in Fig. 8-34, adjust the height of P-3.
3. When the envelope is adjusted properly, it will appear as shown in Fig. 8-35.

Head Amplifier Output Adjustment

1. Video and audio inputs should be at zero level—that is, no signal input.
2. Make a recording of an internal signal and play it back.
3. Connect the scope to the head amplifier output test point.
4. Adjust the mix volume control so that channel A and channel B outputs are equal. Equalize A and B in Fig. 8-36.

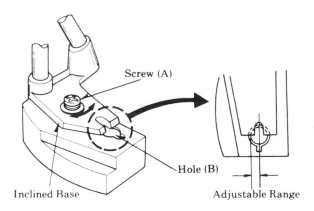

Fig. 8-31 Adjustment of VHS inclined base (*Courtesy,* Sylvania)

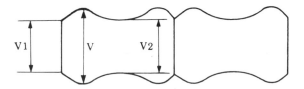

Fig. 8-32 RF envelope adjustment at VHS head amplifier output (*Courtesy,* Sylvania)

Dropping Envelope Level
at End Track

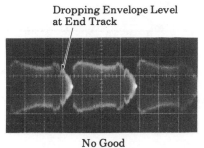

No Good

Fig. 8-33 VHS waveform when P-2 is misadjusted (*Courtesy*, Sylvania)

Dropping Envelope Level
at Beginning of Track

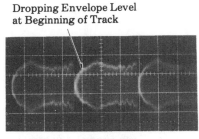

No Good

Fig. 8-34 VHS waveform when P-3 is misadjusted (*Courtesy*, Sylvania)

Envelope is adjusted properly

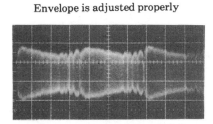

Good

Fig. 8-35 VHS waveform when P-2 and P-3 are properly adjusted (*Courtesy*, Sylvania)

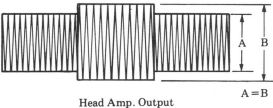

Head Amp. Output

A = B

Fig. 8-36 VHS video head amplifier output adjustment (*Courtesy*, Sylvania)

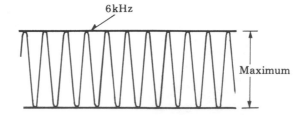

Fig. 8-37 VHS audio output for audio/control head position adjustment (*Courtesy*, Sylvania)

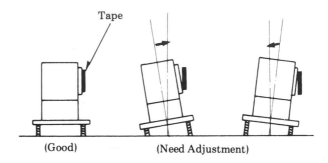

Fig. 8-38 Position of tape with tilted and untilted VHS audio/control head (*Courtesy*, Sylvania)

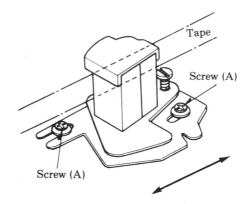

Fig. 8-39 Horizontal adjustment of VHS audio/control head (*Courtesy*, Sylvania)

Audio/Control Head Position Adjustment

Be sure that the tracking control on the top panel is set to the center position.

A/C Head Height Adjustment

1. Connect the scope to the audio output terminals of the recorder.
2. Play back the monoscope portion (6-kHz audio) of the alignment tape (see Fig. 8-37).

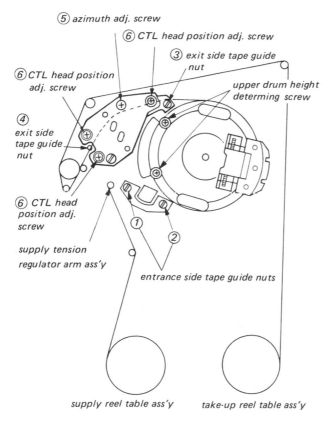

(5) azimuth adj. screw

(6) CTL head position adj. screw

(3) exit side tape guide nut

(6) CTL head position adj. screw

upper drum height determing screw

(4) exit side tape guide nut

(6) CTL head position adj. screw

supply tension regulator arm ass'y

(1)

(2)

entrance side tape guide nuts

supply reel table ass'y   take-up reel table ass'y

Fig. 8-40 Betamax tape path adjustment (*Courtesy*, Sony Corp. of America)

3. Adjust the two screws (A) and screw (B) in Fig. 8-29 for maximum output.

A/C Head Tilt Adjustment

While playing back the monoscope portion of the alignment tape, adjust hex screws (A) in Fig. 8-29 for the maximum RF envelope on the scope. Figure 8-38 shows the conditions of tilt that require this adjustment.

A/C Head Horizontal Position Adjustment

While playing back the monoscope portion of the alignment tape, loosen the two screws (A) in Fig. 8-39, and move the A/C head horizontally for the maximum RF envelope on the scope.

Tape Path (Play) Adjustment (Betamax)

The tape interchangeability adjustments given in the preceding paragraphs for VHS recorders are referred to as Tape Path (Play) Adjustment in Betamax service manuals, and the procedures are very similar. However, because of the different tape loading arrangements and the different tape path of these formats, Betamax recorders require

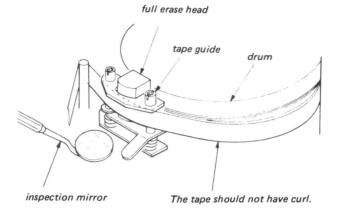

Fig. 8-41 Betamax tape running in its path (*Courtesy,* Sony Corp. of America)

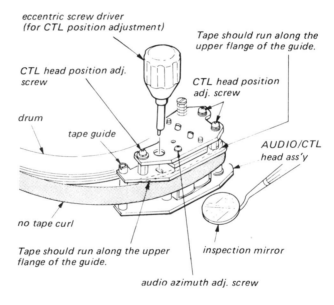

Fig. 8-42 Betamax audio azimuth adjustment (*Courtesy,* Sony Corp. of America)

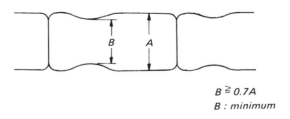

$$B \geqq 0.7A$$
$$B : minimum$$

Fig. 8-43 RF waveform envelope at Betamax video head amplifier output (*Courtesy,* Sony Corp. of America)

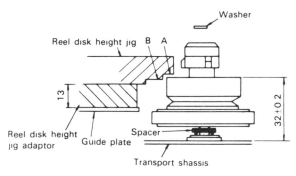

Fig. 8-44 VHS reel table height adjustment (*Courtesy,* Hitachi Sales Corp. of America)

Fig. 8-45 Inserting a spindle height gauge (*Courtesy,* Tentel Corp.)

adjustments at different points, as shown in Fig. 8-40. The equivalent post positions are controlled by tape guide nuts (1), (2), (3), and (4). These nuts are also adjusted to obtain the desired output of the head amplifier. As shown in Fig. 8-43, the results to be achieved are very similar to those specified for VHS recorders. Figures 8-41 and 8-42 show where the tape should not curl. The main purpose of Fig. 8-42, however, is to show audio azimuth adjustment.

Reel Table or Disk Height Adjustment (VHS)

Another adjustment that affects the entry of the tape into the correct path after leaving the cassette is that of reel table or disk height. As shown in Fig. 8-44, jigs and adaptors are used to check the effect of changing spacers of varying thickness at the base of the reel table. The following steps are taken:

1. Check the reel table height using a reel table height jig. The height is between A and B.
2. When the height does not satisfy the conditions in step 1, adjust the spacer and check it again (thickness of spacers: 0.25 mm and 0.5 mm).
3. Fix the washer.

The reel table or disk height adjustment described above requires the removal of the rear panel, top panel, cassette guide, and cassette compartment of the recorder. Tentel has developed spindle height gauges that don't require disassembly of the machine. These gauges are inserted in the cassette compartment, and readings are observed through the "tape viewing" window. Figure 8-45 shows the insertion of such a gauge into a professional recorder.

Spindle height errors are probably the greatest cause of tape edge damage and binding cassette problems with their subsequent effect on tape tension. For further information on spindle height gauges (as well as tension gauges) that are suitable for Beta and VHS home video recorders, contact Tentel.

*Tentel and Tentelometer are registered trademarks of Tentel Corporation, 50 Curtner Avenue, Campbell, California 95008.

9

Troubleshooting
the Electronic Circuitry

Before extensive circuit checking, it is always a good idea to try to localize the trouble to as few sections of the recorder as possible. First, it is easy to determine whether a fault occurs only in the record mode; one merely plays a tape of known quality. If that proves to be OK, the fault must lie in the record circuits of the machine. If it doesn't play back a good recording, then one must look for the fault in the playback circuits.

Needless to say, one of the troubleshooter's first steps is to make sure that the power supply section delivers correct voltages. Manufacturers' service manuals show test points for determining whether correct voltages are being delivered.

The troubleshooting guides in Tables 9-1, 9-2, and 9-3 will be of some assistance in localizing faults:

Table 9-1   Luminance

Symptom	Cause
Luminance Record Circuits	
Noisy picture	Low record current
Low contrast	FM modulator deviation too low
Overmodulation noise	FM modulator deviation too great (may be due to defect in AGC block or misadjusted preamplifier)
Distorted picture	Carrier leakage due to unbalanced FM modulator or sync-tip frequency too low
Lack of detail in white areas of the picture	Misadjusted or defective white-clip stage
Luminance Playback Circuits	
Snowy picture with 30-Hz flicker	Defective video head (cleaning of head may clear trouble; otherwise, both heads must be replaced)

Table 9-1   Luminance (cont'd)

Symptom	Cause
Luminance Record Circuits	
Noisy picture	Defect in playback preamps
30-Hz flicker on peak-white portion of picture	Defect in one of the two preamps
Poor resolution	(1) Preamps have poor frequency response (2) Misadjusted noise canceller
Beat pattern	Carrier leakage due to imbalance in limiters or FM demodulators.
Negative picture	Excessive carrier passing through FM demodulator.

Table 9-2   Chrominance

Symptom	Cause
Chrominance Record Circuits	
Changing color levels	Incorrect timing of "burst flag"
Severe color flicker	(1) Improper phase shift of 4.27-Mhz signal in Betamax units (2) Defective color killer
No color	Loss of signals from AFC or APC sections
Chrominance Signal Playback	
Color flicker	(1) Imbalance between video heads (2) Defect in ACC system
Loss of color lock	(3) Defect in AFC system (4) Defect in APC system
Hue must be readjusted in playback	Color subcarrier frequency incorrect
Loss of color following a dropout	Burst ID pulses not switching flip-flop
Beat patterns in color portions of picture	(1) Leakage of 4.27-Mhz signal into output video circuit (2) Carrier leak from FM luminance circuit beating with chroma signals (3) Interference from local AM broadcast station operating near frequency of the color-under signal (629 kHz in VHS and 688 kHz in Betamax)

Table 9-3   Servo Systems

Symptom	Cause
Picture intermittently broken up by a noise band moving vertically through picture	Servo failing to lock in playback
Switching point moves vertically across picture	Servo fails to lock in record mode
Vertical sync recorded early or late	Servo locks at wrong time

Although these tables of trouble symptoms and possible causes can help localize a fault to a definite area of a recorder, they usually cannot pinpoint the exact component to be replaced. Further troubleshooting requires signal tracing and testing of specific circuits.

Since block diagrams are indispensable for showing the elements and functions of video tape recorders, presenting correct waveforms at various points between the blocks for checking by an oscilloscope would appear to be the ideal format for troubleshooting and servicing any type of video recorder. VTR manufacturers, however, do not present their service manuals in this format. Instead, their detailed circuit diagrams indicate the correct waveforms for specific test points and at certain IC pins and transistor elements. Since most of these points are within the functional circuit blocks, waveforms at points between the blocks do not correspond. Hence, the best one can do is to show examples of tests and adjustments recommended by Betamax and VHS manufacturers. Before doing so, however, we will show how signal tracing equipment is used to analyze various sections of video recorders.

Troubleshooting in the Record Mode

Checking the Record AGC Circuit

As indicated in the troubleshooting tables, a possible cause of overmodulation noise in the recorded picture lies in the AGC circuit ahead of the FM modulator. The Sencore VA48 Video Analyzer, described in Chap. 7, provides all the signals necessary to check this stage. First, the AGC circuit is tested with a reference input level of 1 V P-P, which is injected directly to the camera input of the tape deck from the VTR standard jack of the VA48. The AGC control is adjusted for the proper output level by using the bar sweep video pattern. Second, an adjustable output signal may be fed from the VA48 at its drive signals output jack. This voltage may be varied from 0.5 to 2 V (negative polarity) while the output of the AGC is viewed with an oscilloscope. The waveforms shown in Fig. 9-1(A) should be seen if the AGC circuit is working properly. Figure 9-1(B) shows the waveforms resulting from a faulty AGC circuit.

If the AGC is found to be not working, the next step is to determine whether the dc amplifier, the AGC detector, or the AGC circuits themselves are the cause of the malfunction. The VA48 supplies a bias substitute dc voltage that can be injected as shown in Fig. 9-2 and varied to change

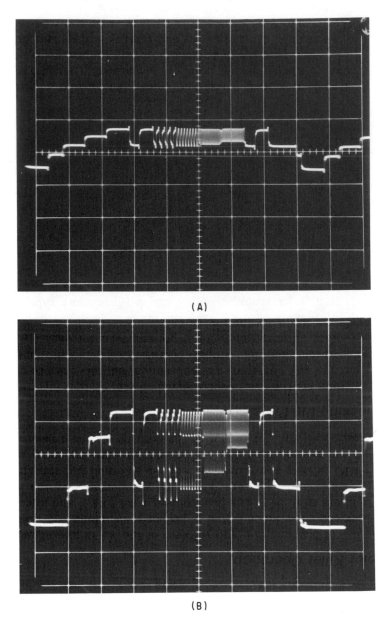

(A)

(B)

Fig. 9-1 (A) Normal, and (B) abnormal (needs troubleshooting) AGC
waveforms (*Courtesy,* Sencore, Inc.)

the output of the AGC stage when a signal is fed to the camera input.
However, if the output does not change, the trouble lies in the gain-
controlled stage. If it does vary, one must go to the AGC detector and
inject the dc voltage at its output. Finally, one can substitute for both
signals feeding the AGC detector itself by injecting the composite video
signal at the camera input (using the VTR standard jack) and then feeding
the composite sync pulses supplied by the drive signal output instead of the
recorder's sync separator output.

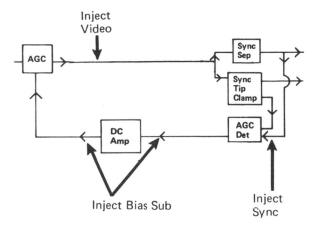

Fig. 9-2 Using drive signals from VA48 analyzer to test each AGC stage (*Courtesy,* Sencore, Inc.)

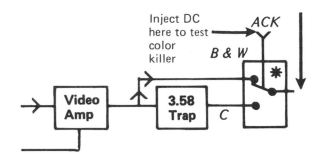

Fig. 9-3 Sweep pattern and bias and B+ subvoltages to check color killer (*Courtesy,* Sencore, Inc.)

Color Killer Operation

The use of bias and B+ sub voltages of the VA48 also allows one to tell whether the color killer signal is properly switching the color/black-and-white filters (Fig. 9-3). The bias voltage is set to 4 V and injected into the IC that controls these filters. The scope waveforms shown in Figs. 9-4 (A) and 9-4(B) indicate the difference between a normal and abnormal color killer operation. Normal operation will allow more high-frequency response during a black-and-white presentation than during a color presentation.

Checking the Pre-Emphasis Circuits

By feeding in the bar sweep signal ahead of the recording pre-emphasis circuits and observing the subsequent output, the pre-emphasis circuits can be checked for proper operation. As a rule, multispeed recorders require different amounts of pre-emphasis for each speed. The scope waveforms following the pre-emphasis circuits will show more high frequency response.

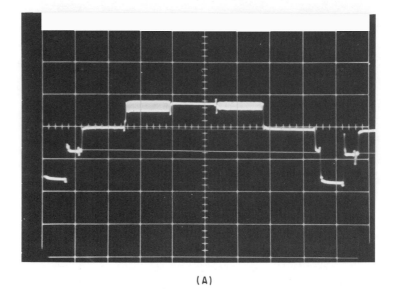

(A)

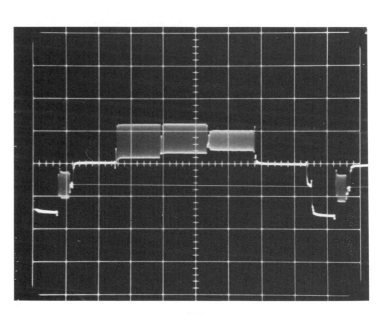

(B)

Fig. 9-4 (A) Normal, and (B) abnormal color killer operation (*Courtesy, Sencore, Inc.*)

Checking White and Dark Clipping

The clipping circuits located between the pre-emphasis network and the FM modulator must be properly adjusted to prevent over-modulation, which will cause the picture to tear during playback. Both a reference white level and a reference black level is required in the adjustment of these circuits to ensure that the delimiters are not favoring one portion of the signal over another. With the three-step grey scale of the bar sweep

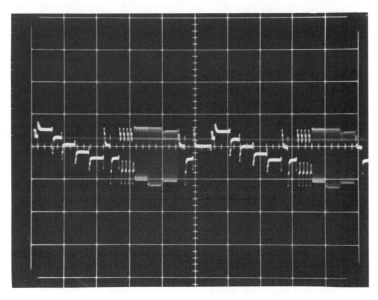

Fig. 9-5 A gray scale and multiburst signal from the bar sweep pattern to allow testing of clipping circuits at all operating frequencies   (*Courtesy,* Sencore, Inc.)

one can check proper video linearity, while the different frequency bars give a dynamic check of the clipping circuits at different video frequencies. The bar sweep pattern allows all frequencies of a signal to be checked at the same time. Thus, one can tell whether clipping circuits are operating properly at pre-emphasized high frequencies as well as low frequencies (see Fig. 9-5).

The circuits from the FM modulator to the video heads may be observed with a scope or the signal tracing meter of the VA48. For general signal tracing, the high frequency response of the meter is actually faster than that of the scope. The meter measures only peak-to-peak amplitude, but this is more important at this point than the shape of the waveforms.

Color Circuit Analyzing

The automatic color control (ACC) requires two input signals for proper operation. The first is the composite chroma signal. The important part of this signal that is needed for ACC operation is the color burst. The amplitude of the burst signal is used to control the gain of the chroma circuits so as to maintain a constant color level with changing input signals. The second signal required is the "burst flag." This signal is simply the horizontal sync pulse that is delayed by a small amount so that its timing comes exactly into line with the burst signal riding on the back porch of the horizontal blanking interval. The timing of this signal is very important because it determines what portion of the color signal will be used to control the gain of the color circuits. If the burst flag arrives too late, the burst gate will separate the first part of the picture (just after the blanking interval) instead of the color burst. The result is that the color levels will be

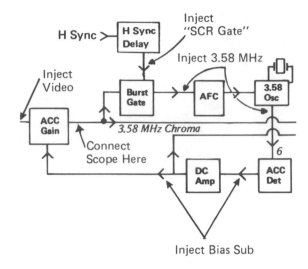

Fig. 9-6 VA48 drive signals used as substitutes for each signal to check the automatic color control circuit (*Courtesy,* Sencore, Inc.)

constantly changing because the amount of chroma information will be different in each color scene.

To analyze a typical symptom of changing color levels when a tape is being played, one needs to know if the levels are changing during the recording or during the playback of the color program. The answer is easily confirmed by simply playing back a tape previously recorded on a machine suspected to be bad on another machine that is known to be working properly. If the color levels remain the same, we know that the defect lies in the playback circuits. If the levels change when the signal has been recorded on the suspected recorder and played back on the reference machine, we know that the trouble lies in the recording circuits of the machine in question.

Changing color levels could be caused by a defect in any of the seven blocks shown in Fig. 9-6. These blocks make up the automatic color control (ACC). A defect in any of these stages will produce almost the same symptom of changing color levels with different input signal levels. Such defects can range from a defective IC to a poor solder joint. One could look for missing signals or bad waveforms with a scope, but the VA48 provides substitute signals to permit a more positive check. One can duplicate the signals that should be produced at the output of each and every stage.

In using the VA48 for servicing the ACC circuitry, the first step is to provide a reference signal at the input to these stages by connecting the VTR standard output of the VA48 to the camera input of the recorder. Following this step, a chroma bar sweep pattern is selected to provide a reference color pattern for the remainder of the analyzing.

The circuits that produce the burst flag signal are checked first. The VA48 Analyzer provides a substitute signal for the normal input to this

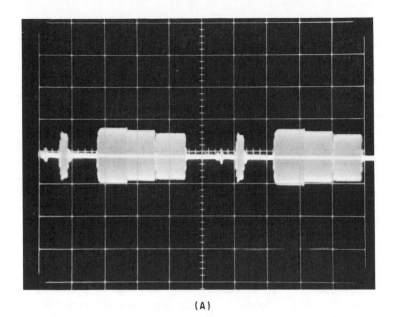

(A)

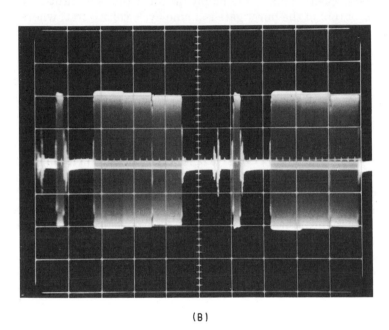

(B)

Fig. 9-7 (A) Normal, and (B) abnormal automatic color control output waveforms (*Courtesy,* Sencore, Inc.)

stage. In the V & H comp sync position of the drive signal switch of the VA48, a substitute for the composite sync pulses separated from the luminance signal is provided. These signals from the VA48 should be monitored by the meter system of the analyzer in the drive signals monitor position to ensure that excessive signal levels will not be fed to the solid-state devices in the circuits being tested.

One starts these tests by injecting the composite sync signal into the horizontal sync delay circuit of the recorder to see if proper operation occurs. The scope waveform shown in Fig. 9-7(A) shows the proper amplitudes for both Beta and VHS formats. Figure 9-7(B) shows the waveform obtained when the ACC circuits are not operating properly. These can be monitored at the output of the ACC controlling stage.

If the substitute signal from the VA48 does not return the proper amplitude at the output of the ACC circuit, one can advance one stage and provide a substitute for the burst flag. In this case, the VA48 drive signals switch is set to provide the SCR gate drive signal. It provides a proper substitute for the burst flag, because the pulse produced by the SCR Gate signal is "stretched" by the same amount as the burst flag. This pulse is present during the color burst and will operate the burst gate just the same as the signal produced by the circuits inside the recorder's chroma-processing stages. If the operation of the ACC circuits now return to normal, the trouble being sought lies in the horizontal sync delay stage.

If these steps have not localized the fault, stage-by-stage injection at the output of the burst gate can be continued. Now one can use the 3.58-MHz (phase-locked) signal. One might question this substitution because the normal output of the burst gate is a nine-cycle color burst, whereas that of the VA48 is a continuous-wave 3.58-MHz signal. However, the AFC circuit (at the output of the burst gate) cannot tell the difference between the color burst and a continuous wave signal and will operate equally well with the latter. Under these conditions, the dynamic operation of the ACC circuit can be checked simply by varying the amplitude of the substitute 3.58-MHz signal.

The output of the ACC detector is a dc voltage whose amplitude is related to the amplitude of the burst signal. The use of bias and B+ sub output of the VA48 allows one to check the operation of both the dc amplifier and the ACC controlled stage. One should start by setting the output of the bias and B+ sub to the level shown on the schematic in the VTR service manual. This voltage is then lowered and raised about 10 percent. There should be a change in the ACC output signal as a result of this lowering and raising of the voltage. If one is still unable to realize proper operation of the ACC, dc voltage should be injected at the output of the dc amplifier (input to the ACC controlled stage) and the voltage again varied. If a changing output level is not realized at this time, it is evident that the trouble lies in the ACC controlled stage.

Although the seven circuits have now been analyzed, the procedure may have looked like the "long way around" the signal path. Actually, the time could be shortened considerably by starting at the output of the ACC detector and supplying the substitute dc voltage. This approach divides the circuits in half. If one obtains proper operation from this point on, it is obvious that the defect is somewhere in front of this stage. If we do not get proper operation, the trouble is seen to lie either in the dc amplifier or the ACC controlled stage.

Checking the Color-Phase Sync Reference and Color Conversion PLL

In addition to the VA48 Analyzer and an oscilloscope, a frequency counter is needed to analyze the color-phase sync reference and color

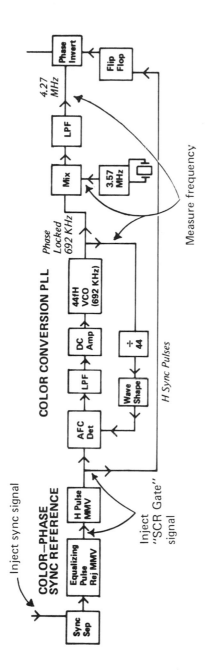

Fig. 9-8 Using substitute signals to check color-phase sync reference and color conversion PLL circuits (*Courtesy,* Sencore, Inc.)

conversion PLL circuitry shown in the block diagram of Fig. 9-8. Basically, the frequency conversion stages mix the incoming 3.58-Mhz chroma signal with a second signal referenced back to the horizontal sync pulses through a phase-locked loop arrangement. The block diagram of Fig. 9-8 is that of a Betamax unit. The operations of such conversion in VHS recorders is similar, although not identical.

In the frequency conversion of this section, the horizontal sync pulses are separated from the incoming composite video signal. These pulses are then formed into a series of pulses with a fixed amplitude (and pulse width) in two multivibrator stages. These clean pulses are then fed to the phase-locked loop to maintain the proper conversion frequency at the output.

The composite sync signals provided by the V & H comp sync output of the VA48 can be fed directly to the input or output of the sync separator stage. These pulses, being phase-locked to the composite video, will then replace the signals that should be at these two points. Such signals, if taken beyond the "equalizing pulse rejection multivibrator," will result in the wrong frequency at the output of the PLL. The reason for this error is that the PLL will try to lock up to the vertical sync pulses (as well as the horizontal sync pulses) and change frequency during each of them. One needs to remember that the function of the "equalizing pulse rejection multivibrator" is to provide a constant pulse rate during the vertical blanking and vertical sync pulse intervals.

This error can be avoided by using the horizontal output (SCR gate) signal for injection after the multivibrator stages. This signal works well because it does not contain the vertical sync pulses but is merely a series of pulses that are phase-locked to the horizontal sync pulses. Since it is an exact duplicate of the output of the "horizontal pulse MMV," it can be injected directly into the AFC detector, the latter being used to keep the PLL output frequency an exact multiple of the horizontal frequency. One point to remember in using the VA48 to supply signals is that they must be of the same polarity and amplitude as the signals normally found in the circuit. Information regarding the signals normally found in the circuit may be found in most VTR service manuals.

The total operation of the PLL is determined by checking the output frequency with a frequency counter. The PLL output frequency should be exactly 44 times the horizontal sync pulse frequency of 15,734 Hz, or 692,296 Hz. If this frequency is not correct, one will find that the recorder will record and reproduce color but that a color tape that has been recorded on another machine will not play back in color. If we do not obtain the proper frequency at the output of the PLL, we simply move the frequency counter to the output of the $\div 44$ stage. At this point, one should find the horizontal sync frequency of 15,734 Hz. If this stage is divided properly, the trouble could be in the low-pass filter or the dc amplifier. The bias and B+ sub supply can be connected to the output of these stages to see if the adjustment of the bias voltage changes the frequency of the PLL output. If there is no change in output frequency when the bias voltage is changed, we know that the defect is in the voltage controlled oscillator (VCO) and that we have a defective IC. Here again, the use of signal substitution

(A)

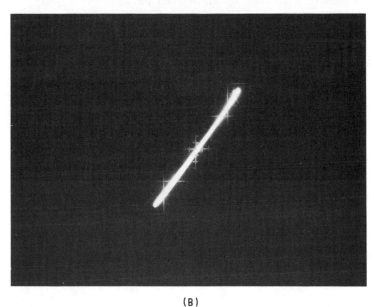

(B)

Fig. 9-9 Using vector mode to test for proper phase shifting of the chroma conversion frequency: (A) Proper operation (normal line-by-line phase reversal), and (B) improper operation (no line-by-line phase reversal) (*Courtesy,* Sencore, Inc.)

allows us to locate a defect with a stage-by-stage analysis. As soon as the proper output frequency is obtained (with a substitute signal injected), we know that the injection point comes after the defective stage.

The operation of the remainder of the frequency conversion stages is simply analyzed with a frequency counter or scope. The second conversion frequency oscillator (a 3.57-MHz crystal-controlled oscillator in the Beta format and a 3.58-MHz crystal-controlled oscillator in the VHS format) is simply adjusted until the proper conversion frequency is obtained. The output of the mixer stage should measure 4.27 MHz in the Beta format and 4.2 MHz in the VHS format. A frequency counter is used for measuring these values.

The vector mode on Sencore's PS163 scope allows one to check for proper phase shifts in the frequency conversion stages. All one needs to do is connect the "A" channel to the 4.27-MHz signal before it is phase-inverted and the "B" channel to the output of the phase-inversion stage. The scope waveforms shown in Fig. 9-9 indicate the patterns that will be obtained if the stages process the phase properly. If they do not, the color will be noisy or there will be no color in the playback mode.

Checking the Color Frequency Response

The chroma bar sweep signals of the VA48 Analyzer provide for a truly dynamic check of the entire color frequency response necessary for a good color picture with color detail in even the smallest objects on the TV screen. The three bars of the chroma bar sweep represent the color subcarrier and the points 50 kHz above and below the subcarrier frequency. Each of the bars is generated at the same amplitude so any can be used as a reference for the total system's frequency response. All one needs to do is to trace the converted chroma bar sweep through the amplifier stages to make sure that color detail is not being lost during the recording process. The waveform in Fig. 9-10 shows how chroma bar sweep patterns are processed in a properly operating recorder. Although there is some loss of high frequency color detail, this is normal. These patterns from the

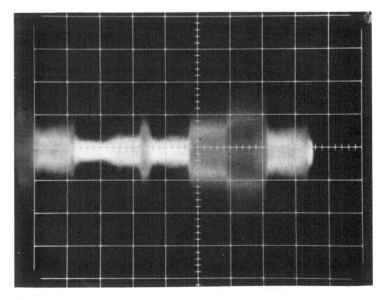

Fig. 9-10 Full chroma bandwidth of the color circuits checked by the chroma bar sweep of the VA48 (*Courtesy,* Sencore, Inc.)

VA48 Analyzer are the only ones that enable the service person to see the amount of high frequency loss present.

Troubleshooting the Playback Circuits

A key element in checking playback circuits is a prerecorded alignment tape or reference tape. Prerecorded alignment tapes are available from the video tape recorder manufacturers, and everyone in the business of servicing VTRs should have one for the VHS and one for the Beta format. However, since these tapes are relatively expensive, it would be a good idea to follow the suggestion in chap. 7 and record your own alignment or reference tape to supplement the one from the manufacturer. If one has a VA48 Analyzer, the easiest way is to record a tape from it (see Chap. 7 for the approach to take). You can then use your own reference tape for most of your work and save the alignment tape from the factory for those occasions where tests and adjustments must be made according to the service manuals available from the VTR manufacturers.

Checking Video Frequency Response in the Playback Mode

If one has the bar sweep pattern on a reference tape, it is necessary only to play this tape and look at output of the machine with a scope to check the video frequency response in that mode. The scope waveform in Fig. 9-11(A) shows the output of a properly operating unit. Even in this case, it will be noted that the high frequency response falls off above 3 MHz. If it drops off above 3 MHz as shown in Figs. 9-11(B) and 9-11(C), the response is being lost at some point in the circuitry of the playback mode. A likely place is in the video head equalization circuits, which compensate for their nonlinear output. In Fig. 9-11(B), high-frequency details are reduced for both the channel A and channel B pickup head. In Fig. 9-11(C), channel A has an overpeaked midband response while channel B does not, resulting in the extra peaking on the lower frequency bars. Manufacturers' service manuals usually outline a recommended method for checking head equalization with their alignment tape. This method should be followed to check video head equalization. (Also see Chap. 8.)

The chroma bar sweep pattern makes it possible to check for good color detail at both the upper and lower frequency limits. The center bar at 3.56 MHz provides a reference level for a comparison of the frequency detail at 500 kHz above and below the subcarrier. Another important point about all three of the bars produced by the chroma bar sweep is that they are phase-locked to the horizontal sync pulse and have a 180-deg phase shift every horizontal line. With these characteristics, they can be properly separated by the comb filters used to cancel color crosstalk during playback. Thus, a means is provided for adjusting the comb filter through the use of the alignment tape, which contains the chroma bar sweep pattern (see Fig. 9-12).

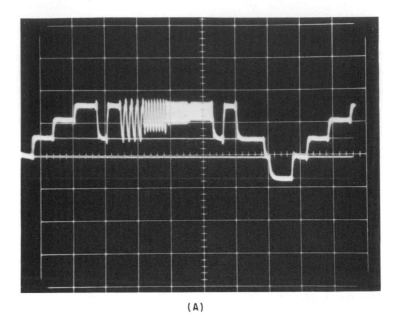

(A)

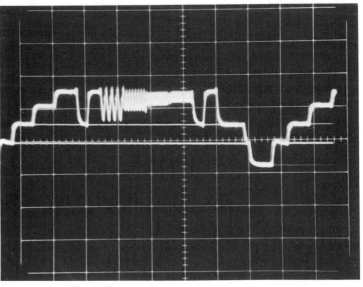

(B)

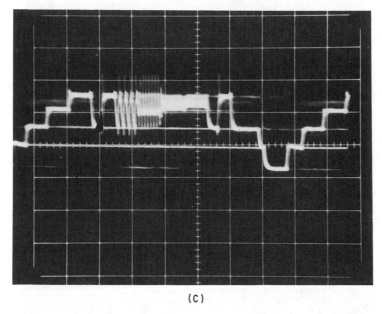

(C)

Fig. 9-11 Frequency response of normal playback circuits: (A) With good video head alignment,   (B)   with poor video head alignment (channels A and B equally affected), and   (C)   with poor video head alignment (channel A with overpeaked midband response) (*Courtesy, Sencore, Inc.*)

Adjusting the Comb Filter Used in the Playback Mode

It is easy to set the comb filter used in the playback portion of the color circuits provided one has a scope with good vertical sensitivity in the order of 5 mV. The Sencore model PS163 dual trace oscilloscope meets this requirement when used with a direct probe. One should start by connecting the scope through the direct probe to the output of the comb filter bridge, which is not connected to the chroma amplifiers. This output point will show the crosstalk rather than the chroma output. At this point, since the signal does not have sync pulses, the external trigger input should be connected to the video output jack for triggering. Then the scope should be set to trigger at the horizontal rate. The sync separators of the PS163 scope will provide a stable trace with the composite video signal used as a reference.

The next step is to play back that portion of the alignment tape that contains the chroma bar sweep pattern. Adjust the comb filter's mixer control until the amplitude of the signal contains the least amount of the second bar, as shown in Fig. 9-12(B). The scope will have to be set for maximum sensitivity to obtain this waveform because the signal level is very low at this point. However, only the mixer control should be adjusted in this case, and the two phasing coils of the comb filter should be left in their factory-aligned positions. The adjustment of these coils is referred to as the *phase adjustment*. Some service manuals of VTR manufacturers

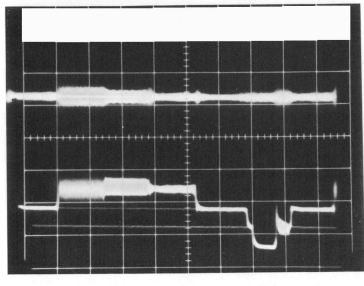

(A)

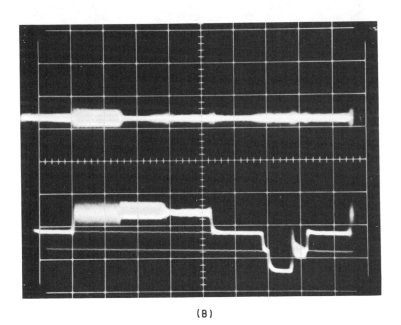

(B)

Fig. 9-12 Adjustment of comb filter: (A) Improper, resulting in greater crosstalk in the playback mode (video output is given for reference), and   (B) proper, giving best crosstalk rejection (video output is given for reference)(*Courtesy,* Sencore, Inc.)

recommend that this adjustment be made only in the factory with a broadcast vectorscope that is able to provide a vector display of the color bars.

Troubleshooting the Video Input System in the Playback Mode

A defective circuit in the video input system will result in a noisy picture with severe flicker. The cause of this symptom is the one-sided effect that only one video input circuit is working properly. The two-head system of VHS and Betamax home video recorders uses one of the heads to pick up every odd field and the other head to pick up every even field. When one of the input signal paths is defective, each complete field is followed by a period of weak signal and noise or noise only.

Each of the video heads has its own rotary transformer, which transfers the signal from the head output to the preamplifier. Each rotary transformer is composed of two coils, one of which is on the moving head disc and the other on the stationary part of the drum assembly. The video head picks up the signal from the tape, after which it is inductively coupled to the stationary coil by the moving coil. The use of rotary transformers eliminates the need for slip rings or brushes, which would surely be a problem with rotary heads. Following each rotary transformer is a switch to go between record and playback. Then there is a preamplifier for each head in the playback mode. Each preamp has a set of adjustments to compensate for any differences in frequency response between the two channels. The signal levels at the input to the preamplifiers should be about 1 mV. If the signal level is low, several points in the above channels could cause the trouble mentioned.

The first possible cause of the defect is the video head itself. It could require cleaning or be worn excessively. Both cleaning and replacement of video heads were treated in Chap. 8. A second possible cause may lie in a rotary transformer. A broken wire leading to the moving coil of one of the rotary transformers would mean that signals from one of the video heads would not reach its preamp. Next, we could have a dirty contact in one of the switches that selects between the record and playback circuits. Again, one of the channels would be getting through. Finally, there is the possibility in these chains of a defect in one of the preamplifiers. To analyze these circuit elements, a signal of a sufficiently low level is required. It should be of the same magnitude as those normally found in the circuits when they are operating. (A somewhat larger signal will result in cross-coupling and make it impossible to identify the defective stage.)

The VA48 Analyzer can provide such a low-level signal at 4.5 MHz when it is used with a 40-dB attenuator, which will bring it down to the 1-mV level. The best place to look for signal output is at the output of the resistive matrix, which is used to mix the output of the two video heads. One must take into account the fact that an electronic switch is used to switch the "A" head output and the "B" head output during playback. This switch is normally controlled by a pulse that comes from servo circuits. When using the substitute signal to analyze the video input circuits, one connects the bias and B+ sub supply instead of the head switching pulse. When the bias voltage is supplied, the electronic switch will switch over to one of the head preamps, and when the bias is removed, it will switch to the other preamp.

Troubleshooting the Drop-Out Compensator

The same output signal from the VA48 used to troubleshoot the video input circuits can be used without the 40-dB attenuator to troubleshoot the drop-out compensator (DOC) circuit in the playback chain. All one needs to do is supply a signal at the input of the DOC detector and look at the output of the DOC circuit with a scope. The DOC circuit is designed to switch to the delay line output any time the signal level from the video heads drops below a certain level. When the 4.5-MHz signal is injected, the DOC detector should switch the signal around the delay line. When the signal level drops below the detector trigger level, the circuit should switch back to the delay line. Since we are no longer feeding a signal into the delay line, the output quickly drops to zero.

The DOC trigger circuit is tested by increasing the RF-IF control to full output and then reducing the signal level. When the level control is at about 10 mV, the output signal should suddenly disappear. Increasing the signal level should then restore the output. If the DOC circuit is not operating properly, one should use a dc voltmeter to check the output of the DOC detector. The detector should provide a dc voltage to control the switching circuits inside the IC. If this voltage change is seen when the signal level is changed at the input, we know that the detector is working properly and that the defect is in the switching circuits. If the voltage does not change, the defect lies in the detector itself. The delay may also be checked by feeding the 4.5-MHz signal to its input and checking for an output.

Checking the Limiter Circuit

A defective limiter will result in a playback signal that varies in detail and noise content. The limiter may be tested by the 4.5-MHz signal using adjustable level. The function of the limiter is to compensate for changing levels in the playback signal so that the FM demodulator always has enough signal to operate properly. All that is required to test the limiter (or limiters, since a double limiter is used in most VHS designs) is to feed the 4.5 MHz into the limiter input and look at the output level. The output level should remain almost the same over a full range of input signal (Fig. 9-13).

Checking Color Processing Circuits in the Playback Mode

The color processing circuits in the playback stages are treated in the same way as those in the recording stages. In fact, several of the same circuits are often used in both modes. The bias and B+ sub output of the VA48 may be used to check any of the automatic circuits by substituting for the feedback voltage if a change occurs in the condition that we are trying to correct. It should be remembered that when the VA48 signals are substituted in the playback circuits they are not phase-locked to the signals

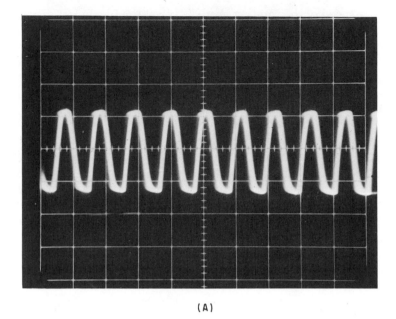

(A)

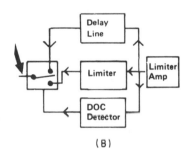

(B)

Fig. 9-13 Checking limiter to determine whether output is constant (*Courtesy,* Sencore, Inc.)

on the prerecorded tape. However, this situation does not prevent the use of substitute signals. They can be injected at the input to a circuit, such as the playback frequency conversion PLL, and the resulting output can be observed to see if the signals are being processed properly.

Troubleshooting the Servo Circuits

The servo control circuits of a video recorder compare two signals to decide whether the video heads are in the correct position during recording and playback. During recording, these two signals are the vertical sync pulses of the TV signal and the head position signal (PG pulse). They should arrive at the servo comparison circuits at the same time. If their timing is not identical, the servo circuits adjust the rotational speed of the video heads until the two signals correspond.

During playback, the same head reference pulse (PG pulse) is compared with the output of the control track head. Then the servo system adjusts the speed of the rotating head disc until the two pulses are properly timed.

Betamax recorders use the servo circuits as a protection circuit as well as a control circuit. The servo logic circuits "look for" a servo pulse coming from the spinning heads. If this pulse is missing, it may indicate that the video head is not spinning because of a jammed tape or a defect in the motor or drive belt that runs the head drum. Any time the PG pulses are missing, the tape deck automatically actuates the "stop" mode to prevent the possibility of further damage.

Signal substitution may be used to analyze many sections of the servo system, which are difficult to analyze using a scope or other servicing technique. A typical tape deck symptom is that the unit will not stay in the "play" mode. The machine will run about 1 sec after the play button is pressed and then automatically shut down. One cannot use an oscilloscope to analyze this defect because the signal produced by the PG head is present only when the video head is spinning. Of course, since the head will spin only when the deck is in the play mode, there is no signal to trace. Consequently, one needs to substitute for the PG signal, which would be coming from its pickup heads. This signal is normally a 30-Hz pulse, but the circuits will also accept a 60-Hz pulse because the PG input circuits are multivibrators. Such a signal is available from the drive signals output of the VA48 Analyzer. It can be adjusted to a 1-V peak-to-peak level and fed into the input of the servo board in place of the PG head output. If the tape deck now runs in the play mode, it becomes apparent that the defect is a missing PG pulse. Its absence may be caused by a defective PG pickup head, or the wiring might be open between the head and the servo board.

If the substitute signal does not cause the unit to operate continuously in the playback mode, it is left connected and signals are traced with a scope. The advantage of this technique is that the tape deck does not have to be running to permit continuation of the signal tracing. Since the PG pulse is simulated by the 60-Hz signal from the VA48, any signals that result from the presence of the PG pulse will appear at the proper points. It is only necessary to remember that the waveforms are going to be slightly different from those shown in VTR service manuals for a 30-Hz PG signal. Typical waveforms to be expected with the substitute signal are shown in Fig. 9-14(b).

The other signal that operates the servo circuits during playback in multispeed models is generated by an internal 30-Hz generator. In this case, a 300-Hz oscillator is phase-locked to the ac line. The 300-Hz signal is then divided by 10 to obtain a 30-Hz signal, which is used as a reference to compare with the 30-Hz control track signal. This section of the servo circuits is analyzed by signal tracing with the VA48's built-in peak-to-peak meter, a frequency counter, or a scope. The use of the PR50 audio prescaler with the Sencore FC45 frequency counter provides two additional digits of resolution (to 0.01 Hz) for accurately troubleshooting this section.

Electrical Alignment

If one is to attempt electrical alignment of either Betamax or VHS recorders, it is essential that a service manual for the specific model of

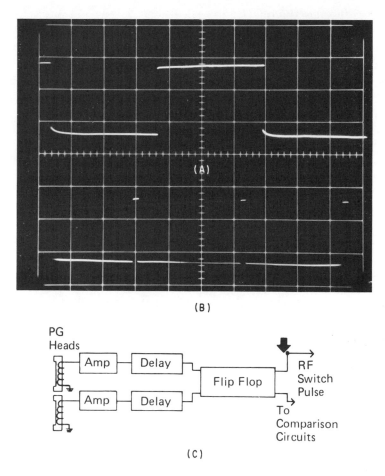

Fig. 9-14 (A) Normal pulse generator, with signal at point indicated; (B) signal being substituted for (A); and (C) block diagram of circuit from PG heads to RF switch pulse (*Courtesy,* Sencore, Inc.)

recorder be available for reference in regard to test points, integrated circuit pin numbers, and designated transistor elements where measurements are to be made.

Video System Alignment

For Betamax units, it is recommended that the playback system be aligned first with an alignment tape supplying the input signal. The record system is aligned after it is confirmed that the playback system is operating properly. The Y-signal and chroma-signal systems are aligned for each of the playback and record systems. The color video signal, supplied from the tuner block, is used as the video input signal in the video system alignment. A check must be made to see if the sync and burst signal of the video input satisfy the required specification.

It is necessary that the following alignment adjustments and checks be made in the exact sequence shown:

Playback System Alignment

1. Playback RF amplifier frequency response adjustment
2. Dropout compensator threshold level adjustment
3. Playback video output level adjustment
4. 44fH VCO oscillating frequency adjustment
5. 3.57-MHz VXO free-running frequency adjustment
6. 3.58-MHz oscillator frequency adjustment
7. 4.27-MHz carrier phase alternate check
8. 4.27-MHz carrier leak adjustment
9. Playback color killer adjustment
10. Playback chroma output level adjustment
11. Comb filter adjustment

Record System Alignment

12. Sync-clip carrier set and dark-clip adjustment
13. Carrier balance adjustment
14. 3.58-MHz trap adjustment
15. FM modulator deviation adjustment
16. E-E output level adjustment
17. 3.58-MHz crystal filter adjustment
18. Record ACC level adjustment
19. Y-FM signal record current adjustment and chroma record current check

Drum Servo and Pulse System Alignment

1. Drum free speed check
2. DC amplifier bias adjustment
3. RF switching position adjustment
4. Record-servo lock-phase adjustment
5. Playback CTL signal check
6. Tracking control set
7. Play-servo phase adjustment

Audio System Adjustment

1. Audio head azimuth adjustment
2. Playback frequency characteristic adjustment
3. Playback output level adjustment
4. Bias oscillator check
5. Bias trap adjustment
6. Record bias adjustment
7. Record current adjustment
8. Overall frequency characteristic check
9. AGC operation check

10. S/N ratio check
11. Distortion check

Typical VHS service manuals list the following electrical adjustment procedures for the video, servo, and audio sections of VHS recorders:

Video Section

1. Head amplifier peak frequency adjustment
2. Head amplifier frequency response and balance adjustment
3. E-E level adjustment
4. Sync-Tip frequency and deviation adjustment
5. White-and dark-clip adjustments
6. Recording current adjustment
7. Dropout detector input level adjustment
8. Limiter balance adjustment
9. 3.58-MHz crystal oscillator adjustment
10. LP/SP burst level confirmation
11. 1.5-µs pulse adjustment
12. AFC adjustment
13. APC 3.58-MHz VXO adjustment
14. Comb filter adjustment
15. Balance modulator adjustment
16. Color killer adjustment
17. Video/chroma level adjustment
18. Audio bias trap adjustment

Servo Section

Adjustments in stop or playback modes

1. 60-Hz oscillation confirmation
2. Switching flip-flop duty cycle adjustment
3. Control head output confirmation
4. Tracking fix adjustment
5. P.G. head output confirmation
6. Oscillation amplitude adjustment
7. Capstan FG output confirmation

Adjustments in the record mode

8. Cylinder servo sampling gate adjustment
9. Capstan servo sampling gate adjustment
10. Head switching position adjustment
11. Control duty adjustment
12. LP/SP voltage confirmation
13. LP/SP switching confirmation

Audio Section

Adjustment in record mode

1. AGC confirmation
2. Bias current/dummy coil adjustments

Playback gain adjustment

10

Personal Video Camera Theory and Servicing

When home and personal portable video cassette recorders were first introduced, a number of monochrome TV cameras were sold that are still in the hands of users. Most owners of such cameras would undoubtedly have preferred color cameras had such units been available at affordable prices. Development efforts of TV camera manufacturers directed at color units in the $1000 to $1500 price range, or less, have been successful in providing color TV cameras capable of very satisfactory performance considering their price range. Since public interest now lies mainly in color units, this chapter will discuss color TV cameras first and conclude with monochrome units.

Broadcast Color TV Cameras

Color TV cameras capable of the highest quality output ordinarily use three camera tubes. For broadcasting or electronic news gathering, these tubes are arranged in the configurations shown in Figs. 10-1(A) and (B). The colors are separated prior to the entry of the light into the pickup tubes by either prisms or dichroic mirrors with the result that each pickup tube

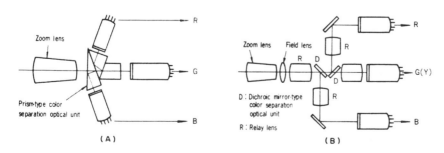

Fig. 10-1 Typical configurations of three-tube broadcast color TV cameras (*Courtesy, Journal of Electronic Engineering,* Japan, K. Wakui, NHK)

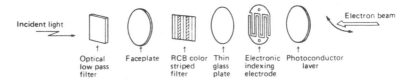

Fig. 10-2 Target components of Sony Trinicon camera tube (*Courtesy, Journal of Electronic Engineering,* Japan, K. Yamagata, Sony)

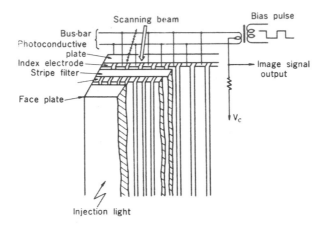

Fig. 10-3 Cross section of the Trinicon target structure(*Courtesy, Journal of Electronic Engineering,* Japan, Wakui, NHK)

delivers a separate color. In the NTSC system, these colors include red, green, and blue. The luminance (Y) signal may be derived from the green signal as shown in Fig. 10-1(B). A more idealized camera would use four pickup tubes—three for the colors and the fourth for the luminance signal—but this arrangement is seldom employed.

The cameras in Fig. 10-1 usually employ plumbicon or saticon pickup tubes. Both have excellent spectral sensitivity, very good resolving power, and a high response speed, even in sizes as small as ⅔ in. in diameter. Hence, they are highly suitable for incorporation into three-tube configurations of reasonable size. Since cameras of this class range in price from $20,000 to $100,000, they bear little relationship to those designed to sell in the $1000 to $1500 range. Nevertheless, basic tube developments such as plumbicon and saticon can sometimes be designed in a different form to handle all three colors in a single tube at a much lower price.

Single Tubes for Color TV Cameras

Although three-tube designs are being used extensively in broadcast color TV cameras, only the development of single-tube color TV cameras has made it possible for home and personal portable cameras to become available at reasonable prices. This single pickup tube must be capable of delivering the luminance (Y) signal and the color difference signals (B–Y) and (R–Y) that are required to form an NTSC color signal. Single tubes for color cameras fall into three general types, as follows:

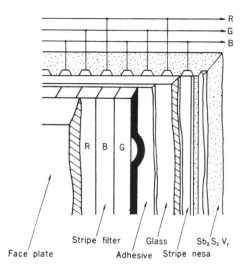

Fig. 10-4 Cross section of three-color Vidicon target (*Courtesy, Journal of Electronic Engineering,* Japan, K. Wakui, NHK)

1. Dot sequential with time multiplex processing of an encoded single output from the pickup tube
2. Dot sequential with frequency multiplex processing of an encoded output from the pickup tube
3. Frequency separation providing three colors separately at the output of the pickup tube

All these tapes employ a color striped filter. In addition, the first and second types employ an electronic indexing electrode between the striped filter and the photoconductive layer, for instance, the Sony Trinicon, an example of the first type (see Fig. 10-2). The third type delivers red, green, and blue signals on separate output leads and does not employ an indexing electrode. Such tubes are usually vidicons, although the single-tube saticon also fits into this class.

A comparison between the cross-section of the target of the Sony Trinicon and that of the three-color frequency-separation vidicon may be made in Figs. 10-3 and 10-4.

Since the first two types of tubes listed do not deliver red, green, and blue signals on separate output leads, it is necessary to achieve effective color separation in the circuitry following the camera tubes, or at least to provide signal separation that results in Y, B–Y, and R–Y signals at the camera output. Such basic circuitry is shown for the Trinicon in the block diagrams of Fig. 10-5. A signal passing through a trap at 4.5 MHz is used to establish the Y output, whereas a signal passing through a bandpass filter centered at 4.5 MHz is used to establish the chroma signals needed to produce B–Y and R–Y output. Comb filters may also be used to decode the signals, as shown in the basic block diagram of Fig. 10-5(B).

The circuitry for the third type—in which the red, green, and blue signals are handled separately—delivers the three color signals through

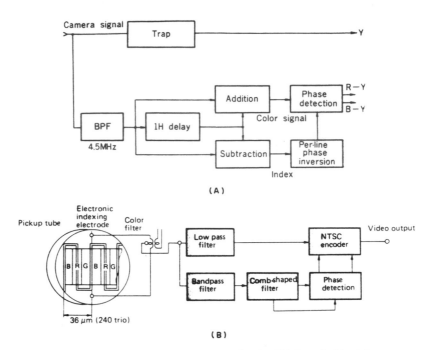

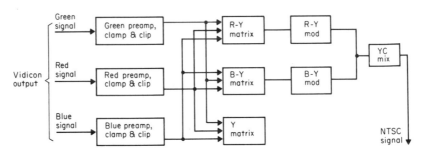

Fig. 10-5 Two circuit arrangements for the Trinicon in block form (*Courtesy, Journal of Electronic Engineering,* Japan)

Fig. 10-6 Block diagram of circuitry for three-color vidicon

separate preamplifiers, clamps, and clippers to the inputs of the matrix blocks, as shown in Fig. 10-6. The R–Y, B–Y, and Y signals are developed in the matrix blocks, and the first two are then fed through low-pass filters and clamps to modulators and final mixing with the Y signal. The resulting output is an NTSC signal of the proper mix.

Personal Color TV Cameras

All three pickup tubes just discussed will be found in the personal color TV cameras currently on the market. We will now provide a detailed description of the circuitry and adjustment procedures for a camera of the third type, our material being based on Zenith's models KC-1000 and

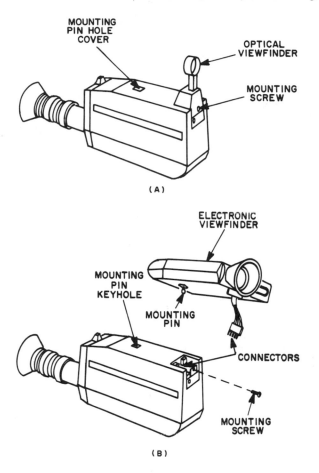

Fig. 10-7 Video camera viewfinders: (A) Optical, and (B) electronic (*Courtesy,* Zenith Radio Corp.)

KC-1250 and presented here with the permission of the Zenith Radio Corporation. The cameras of these two models almost are identical, differing only in that the KC-1000 is fitted with an optical viewfinder and the KC-1250 with an electronic viewfinder. These viewfinders attach to the cameras as shown in Fig. 10-7. The video, timing, and sweep circuits are described in some detail and also shown in block diagram form. Troubleshooting charts are included for the camera as well for the electronic viewfinder. All these methods may be used to service a wide range of color TV cameras and electronic viewfinders since they are not specifically limited to these Zenith models.

The illumination test setup and the camera adjustment setup shown in Figs. 10-11 and 10-12, respectively, are also broadly applicable to the personal color TV camera field.

The Zenith camera (KC-1250) with electronic viewfinder is shown in Fig. 10-8. Each control, indicator, and connector is indicated by a number, and their functions are described below:

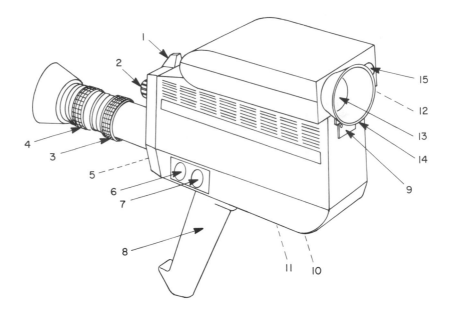

Fig. 10-8 Camera and electronic viewfinder controls, indicators, and connectors: (1) Low light indicator, which glows to indicate power on and adequate light level for camera operation, and flickers to indicate inadequate light level; (2) microphone to record sound close to camera; (3) focus for adjusting lens; (4) lens aperture to control iris opening of the lens; (5) remote pause pushbutton for putting video cassette recorder in pause mode (provided recorder has pause feature); (6) brightness control to set background level; (7) Colortemp control to adjust camera to type of lighting; (8) handgrip (removable) for handheld shots; (9) VTRPAUSE, which lights when video cassette recorder is in pause mode; and (10) EXT. MIC., the connector for the external microphone (*Courtesy,* Zenith Radio Corp.)

Camera

1. Low light indicator: Glows to indicate power on and adequate light level for camera operation; flickers to indicate inadequate light for proper camera operation
2. Microphone: Picks up sound (close to camera)
3. Focus: Adjusts focus of camera lens
4. Lens aperture: Controls iris opening of lens
5. Remote pause pushbutton: Puts video cassette recorder in pause mode, provided recorder has pause feature
6. Brightness: Sets background level
7. Color Temp: Adjusts camera to type of lighting
8. Handgrip: Removable grip for hand-held shots
9. VTR Pause (light): Lights when video cassette recorder is in pause mode
10. Ext. Mic.: Connector for external microphone

11. Camera cable connector: Connector for cable between camera and power supply

Electronic Viewfinder

12. Deck Monitor: Volume control and on-off switch (pull) for video-playback (push for viewfinder)
13. Viewfinder: 1.5-in. CRT indicating camera view
14. Eye cap: Allows better visibility of viewfinder in bright light
15. Speaker: Used for playback from VCR
16. Connector: Electrically connects viewfinder to camera (located under viewfinder)

Camera Adjustments and Settings

Aperture and focus should be adjusted according to the procedure given in the camera operating guide:

1. When the camera is not in use, the aperture should be set to the fully closed position.
2. When in use, if the low-light indicator flickers, open the aperture so that the low-light indicator changes from a flicker to a steady glow.

For normal operation, the brightness control should be set to a center position. Clockwise rotation of this control increases the black level set up, whereas counterclockwise rotation decreases it.

Color Temperature Adjustment

The human eye is able to adapt itself with respect to light in such a way that an object will appear white at various light levels. For instance, if an object appears white under indoor illumination, it will also appear white when viewed outdoors. However, a camera does not possess this adaptability so that an object that is white under certain conditions will appear to have a reddish or bluish hue at other light levels. Color temperature adjustments are performed in order to compensate for this characteristic of the camera. The variations in color temperature for various conditions of lighting are as follows:

Location	Approximate Color Temperature
Indoors with halogen or tungsten lamp illumination	3000°K–3200°K
Indoors with white fluorescent illumination	4500°K
Outdoors illumination (direct sunlight)	6000°K
Outdoors illumination (cloudy; daylight fluorescent)	6800°K–7000°K

The operating guide for the camera will show the correct indicator positions to set the color temperature control.

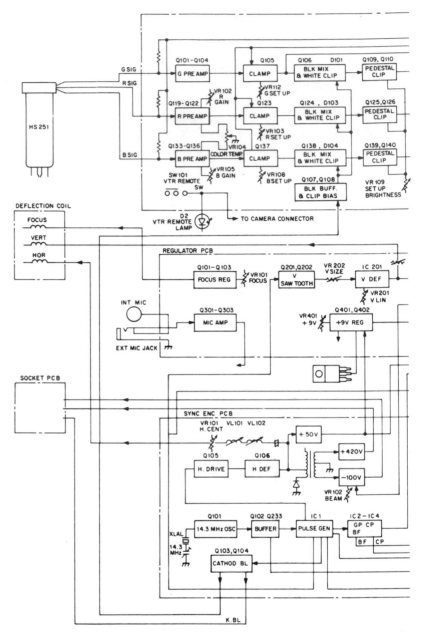

Fig. 10-9 Block diagram of detailed three-color

Video Circuits

The heart of this camera is the tri-electrode vidicon tube. Three electrodes or targets inside the vidicon provide the red, green, and blue colors, respectively. The single vidicon delivers all three colors as separate signals. The circuitry following it is of the general form shown in the block diagram of Fig. 10-6 previously described. Figure 10-9 presents a much

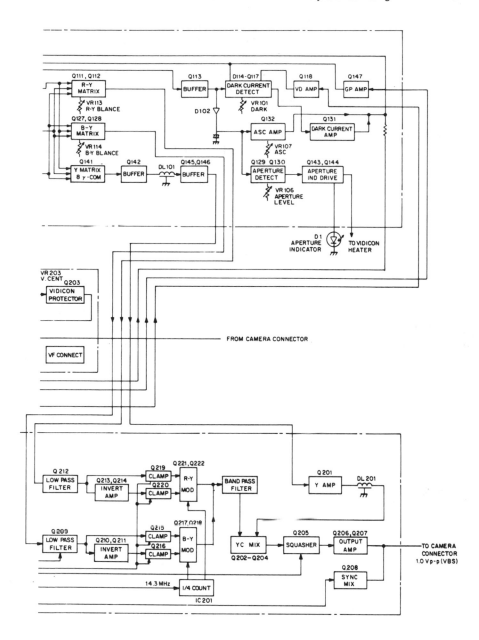

vidicon circuitry (*Courtesy,* Zenith Radio Corp.)

more detailed block diagram of the circuitry, which is described fully below.

The three color signals are applied to separate, corresponding preamplifiers. The gain of the green preamplifier is fixed, but that of the red preamplifier is adjustable. The blue preamplifier contains the color temperature control, which adjusts the balance between the red and blue preamplifiers. This control—available externally—is labeled COLOR

TEMP control. The signal level at the output of the preamplifiers is about 0.8 V peak-to-peak. The optical black level for each of the three color signals is set by gated clamp stages to produce stable black balance. The three clamp stages are gated by a gate pulse from the sync-encoder board. This pulse is generated during each horizontal retrace period ensuring that video is at the black level during retrace. The three color signals are then routed through the blank-mix and white-clip stages. The white level of the video signal is partially affected by the setup control, which determines the signal level at which the diodes conduct and limit signal amplitude. Blanking of the signal occurs during each horizontal retrace period by the application of a blanking pulse from the sync-encoder to the blank buffer, thereby clamping the video at a level below the optical black level. The pedestal clip stages clip the signal to establish the pedestal level. The pedestal level is that at which sync pulses will be added in later circuits to form composite video.

At this point, the three color signals are applied (1) to the R–Y matrix circuit and the B–Y matrix circuit to form the chrominance signal and (2) to the luminance (Y) matrix and gamma correction circuit to form the luminance signal. The R–Y balance control sets the amount of green signal, and this is combined with a fixed ratio of red and blue to form the R–Y signal. The B–Y balance control sets the amount of blue signal, and this is combined with a fixed ratio of red and green signal to form the R–Y signal. The Y signal is formed by combining red, green, and blue in a fixed ratio determined by a resistor network. The amplitude of the Y signal is controlled by the gamma corrector. When the luminance signal increases beyond a preset point, a transistor in the Y matrix conducts and provides a shunt impedance at the input of the following buffer. Thus, the signal strength is reduced. The output of this buffer is applied through the first section of a two-section delay line. This line delays the luminance signal 0.7 μsec to compensate for delays in the chroma circuit and ensure that the final composite video signal will have the correct phase relationships between the R–Y, B–Y, and Y signals. The Y signal is then applied through a buffer to the sync-encoder board where it is amplified, delayed another 0.7 μsec, and applied to the YC (Luminance-Chroma) mixer to form the video portion of the composite video signals. The output of the R–Y and B–Y matrixes are also applied to the sync-encoder board through low-pass filters. The outputs from the filters are applied to balanced modulators. Since both of them operate identically, only the R–Y processing will be described.

The R–Y signal at the output of the low-pass filter is applied directly to the R–Y modulator and clamped. A second output of the low-pass filter is inverted 180 deg out-of-phase with the first signal. The R–Y signal is clamped during retrace to provide dc restoration. The R–Y modulator also receives the 3.58-MHz color subcarrier that is generated by using a 14.3-MHz master oscillator circuit and dividing it by 4:1. The signal modulates the 3.58-MHz carrier to form the R–Y portion of the chroma signal. The use of a balanced modulator causes the subcarrier (3.58-MHz) signal to be cancelled in the output, leaving only the modulation for application to the Y–C mixer through a 0.5-MHz bandpass filter.

The B–Y circuit operates the same except that the subcarrier is shifted 90 deg in an IC, and a burst flag signal is inserted at the low-pass filter in the B–Y mod chain. This burst flag introduces an unbalance at the B–Y modulator inputs and allows the 3.58-MHz burst to be added to the B–Y modulator output. The B–Y and R–Y modulator outputs are added and applied through the Y–C mixer to form the composite signal. The burst is timed to appear on the back porch of the horizontal sync; it is the synchronizing signal for the receiver chroma circuit.

Before leaving the video circuit section of the camera, one should also become familiar with certain special circuits. These are the dark current circuit, the automatic sensitivity control (ASC) circuit, and the aperture detect, or Wink, circuit.

The dark current circuit delivers the current to the red, green, and blue targets in the vidicon. The quiescent dark current (current when the lens aperture is closed) is adjusted by the dark control to a level of 0.02 μA. Since the current is affected by ambient temperature variations, a thermistor is used to reduce these effects. As the ambient temperature changes, the thermistor changes the bias on the dark current amplifier, which in turn changes the current to the three targets in the vidicon, correcting for ambient temperature changes.

The ASC circuit varies the output signal level from the targets of the vidicon to compensate for scene lighting changes. It does so by applying a portion of the green signal voltage through a buffer to the ASC control stage. It is set to provide a 1-V peak-to-peak video output signal from the camera. The ASC circuit maintains this level by reducing the voltage to the vidicon as the average green signal level increases.

The aperture detector, which is composed of two transistors, controls the Level indicator, driving it through an astable multivibrator. At low light levels, the astable multivibrator oscillates at approximately 2 Hz, causing the Level indicator to blink on and off. The aperture level control, labeled Wink, sets the signal level at the first transistor in the aperture detector circuit so that it conducts. When the first transistor conducts, the second transistor is gated on, inhibiting the astable multivibrator and causing the Level indicator to remain constantly illuminated.

Timing and Sweep Circuits

All pulses necessary for gating, timing, and sweep initiation are developed from a master 14.3-MHz clock circuit in the sync encoder. The exact frequency is 14.31818 MHz, but for the sake of brevity we shall refer to it as 14.3 MHz. This output is applied through a buffer to a ¼-countdown circuit and through other buffers to a pulse-generator IC. The ¼-countdown circuit IC is used to generate the 3.58-MHz color subcarrier, as explained in the description of the video circuit. The pulse generator provides the following:

1. A horizontal drive pulse at 15.734 Hz to initiate horizontal sweeps
2. A vertical drive pulse at 59.94 Hz for the vertical sweep
3. A blanking pulse for blanking the vidicon and color amplifiers
4. Composite sync pulses that are mixed with the video signal to form composite video
5. A burst flag pulse that gates the back porch of the horizontal sync pulse to allow the color sync signal to be inserted
6. A clamp pulse to clamp the R–Y and the B–Y input to the balanced modulator during blanking time
7. A gate pulse to gate the color preamplifier signals so as to establish the optical black level

The horizontal drive pulse is applied to the horizontal drive stage to initiate the horizontal sweep output. The flyback transformer in the horizontal circuit is used to develop the necessary dc voltages for the vidicon, which include the negative beam voltage and the positive target voltage. The horizontal sweep signal is controlled by the "Hcent" control, the width control, and the linearity control and is applied to the horizontal deflection coil.

The vertical drive pulse is applied to the vertical sawtooth circuit. The vertical sweep signal size is controlled by the variable resistor labeled "Vsize" in the block diagram of Fig. 10-9. Vertical linearity is controlled by the variable resistor labeled "Vlin," and the vertical centering is controlled by the variable resistor labeled "Vcent." The vertical sweep signal is applied to the vertical deflection coil. A portion of the vertical sweep is applied to the vidicon protection circuit, causing its transistor to conduct and thus complete a path for the negative beam voltage circuit. If a failure occurs that results in the loss of vertical sweep, the vidicon protection transistor switches off, opening the beam voltage divider circuit, and causing the beam voltage to increase and cut off the vidicon beam.

Composite Video Circuits

The luminance (Y) and chroma (R–Y and B–Y) signals are combined in the Y–C mixer in the sync encoder as explained in the description of the video circuit. The video output of the Y–C mixer is applied to the output amplifier through a squasher circuit, which is used to insert a blanking pulse and a burst flag pulse from the pulse generator. The pulse blanks the video during the retrace period by clamping the output of the video output amplifier to the collector voltage of the squasher transistor. The burst flag pulse occurs during the blanking period and removes the blanking for approximately 2 μsec. During this 2-μsec interval, nine cycles of the 3.58-MHz burst that was applied through the B–Y circuit are allowed to pass. This time period, which is coincident with the back porch of the horizontal sync pulse, is used to synchronize the chroma circuits in the color receiver. The composite sync signal containing the horizontal and vertical sync pulses is applied through the sync mixer and added to the video signal, which is then fed out of the camera. The composite video signal now consists of the luminance signal, chroma signals, a 3.58-MHz burst signal,

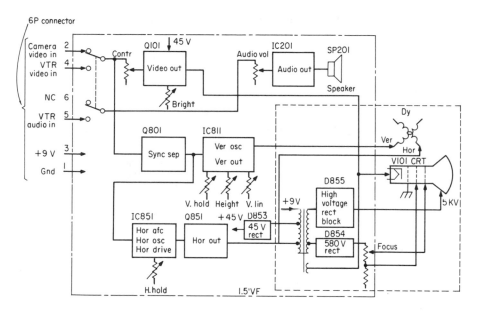

Fig. 10-10 Block diagram of electronic viewfinder (*Courtesy*, Zenith Radio Corp.)

and horizontal and vertical sync pulses. The overall video signal in a properly adjusted camera of this type will be a 1-V peak-to-peak composite consisting of video at a 0.7-V peak-to-peak level plus 0.3-V sync pulses.

The Electronic Viewfinder

The electronic viewfinder contains a small video amplifier, which is fed by signal from the camera. The output of the amplifier is fed to a 1.5-in. cathode ray tube, which displays the camera view in a monochrome picture. Provision is also made to feed a signal from a video cassette recorder into the viewfinder amplifier. A single-chip audio amplifier in the viewfinder will deliver 0.1 W of audio from the video recorder input. A block diagram of an electronic viewfinder is shown in Fig. 10-10.

When a signal is fed from the video cassette recorder to the electronic viewfinder, it will monitor in the record and playback modes. When the recorder is placed in the record mode, the recorded display will be seen in monochrome on the viewfinder. When the playback mode is selected at the VCR, the electronic viewfinder will display the video output of the VCR if it is switched to the monitor mode. The playback audio may also be monitored by another adjustment of the monitor control knob.

Troubleshooting Procedures for Video Cameras

A troubleshooting chart for color video cameras is presented in Table 10-1, and for electronic viewfinders in Table 10-2. *Warning:* Dangerously high voltages exist in video cameras. Use extreme care when servicing.

Table 10-1   Troubleshooting chart for color video cameras

Symptom	Cause
No video output or sound	(1) Power Adapter regulator inoperative (2) Fuse on regulator board open (3) Camera 9-Vdc regulator defective
Composite video present but no audio with audio applied to Ext Mic connection	Failure of audio amplifier circuit
Audio output present when audio is applied to Ext Mic connector but no audio output present when using internal microphone	(1) Internal microphone defective (2) External jack defective (3) Coupling circuits or voltage supply defective
Audio output present but no composite video output	(1) Master oscillator circuit defective (2) No target voltage
Monochrome signal output but no color	(1) 3.58-MHz subcarrier not present (2) Y–C mixer defective
No green color in picture	Failure of green preamplifier
No blue color in picture	Failure of blue preamplifier
Picture reddish in color	(1) B–Y mixer defective (2) B-Y modulator defective
Picture bluish in color	(1) R–Y mixer defective (2) R–Y modulator defective
No red color in picture	Failure in red preamplifier
Picture blurring	(1) Lens not focused (2) Magnetic focus defective
Picture appears granular or with shading produced	Dark current too high
Raster present with faint picture	Low target voltage
Trailing produced with pickup of bright object	Insufficient beam current
Color lock not obtained on TV or VCR	Master oscillator off frequency
Color appears in black portion of picture	Excessive carrier leakage due to improper black balance set up
Color appears in white portions of picture	(1) Color Temp control improperly set (2) Red and/or blue gain incorrect
Aperture indicator blinks even at high light levels with aperture open	Wink circuit defective
Aperture indicator remains On regardless of setting of aperture	Astable multivibrator circuit inoperative; one or more transistors in aperture detector shorted
Aperture indicator does not illuminate	(1) Indicator defective (2) Astable multivibrator inoperative
No raster	(1) Failure of master oscillator (2) Failure of pulse generator (3) Horizontal sweep failure (4) Vertical sweep failure

Table 10-2   Troubleshooting Chart for Electronic Viewfinder
Note: Make sure that input composite video contains both video and sync pulses before troubleshooting

Symptom	Cause
No sound, picture, raster	No operating voltages
No sound; picture present when using viewfinder to monitor playback of VCR	(1) Deck monitor switch in wrong position or defective (2) Audio output IC or speaker defective
No vertical sweep; horizontal sweep present	Vertical sweep circuit defective
Vertical does not synchronize; image rolls vertically	Failure of sync separator or vertical sync input
No horizontal or vertical synchronization	Failure of sync separator
No horizontal sync; vertical syncs properly	Failure in horizontal output transistor
Sound present, raster present, but not picture	Failure of video amplifier
Sound present, no raster	(1) High voltage failure (2) Failure of horizontal sweep circuit (3) Failure of CRT
Retrace lines present	Failure in blanking circuit
Picture cannot be focused and appears dim	Fault in flyback circuit
Picture cannot be focused; brightness appears normal	Fault in focus circuit

It is recommended that the complete adjustment procedure be used whenever the vidicon, coil assembly, or video, regulator and/or sync encoder are repaired or replaced.

Equipment Required

> 600-W tungsten-halogen lamp (3200°K)
> White plastic diffuser (Kolonite plexiglas, #W-2447, P-80 finish)
> Color test chart (DNP Standard color bar chart)
> Gray scale chart (Log gray scale; nine-step; EIAJ; type C1)
> Resolution chart (EIAJ type A)
> Window chart (white window should be 2 in. square; may be made from a sheet of black paper with 2-in. square cut in its center)
> Chart holding frame (any holder or clamp which will hold the diffuser and chart)
> VOM or DC voltmeter
> Audio generator
> Color monitor with video inputs (must be capable of underscan; may be monochrome if video recorder and color TV receiver are used)

Video cassette recorder (not needed if color monitor has video inputs)
Color TV monitor (not needed if color monitor has video inputs)
Oscilloscope (10-MHz bandwidth minimum)
AC voltmeter
Tripod
Frequency counter
58-mm screw-in + 2 close-up lens
Series 6 + 2 close-up lens
37.5-mm series 6 adapter
Light meter (measurements in foot-candles, foot-lamberts, or lux)

Illumination Setup

When testing or adjusting the camera, it is necessary to provide correct levels of illumination and proper light source to insure proper color temperature for adjustment and operation (see Fig. 10-11). The light source is a 600-W 3200°K tungsten-halogen lamp. The four charts used during adjustment procedures are a color bar chart, a gray scale chart, a resolution chart, and a window chart. The gray scale chart must have a transmittance of 0.6 for the white bar. The 600-W tungsten-halogen lamp must be placed 56 in. behind a frame used to hold the charts. Also, a white plastic diffuser sheet must be placed between the light source and the chart to provide the necessary diffusion. The transmitted light level must be 2400 lux (223 foot-candles). If a light meter is available, adjust the chart-to-lamp distance to provide either 223 foot-candles or 223 foot-lamberts or 2400 lux (as read on the light meter).

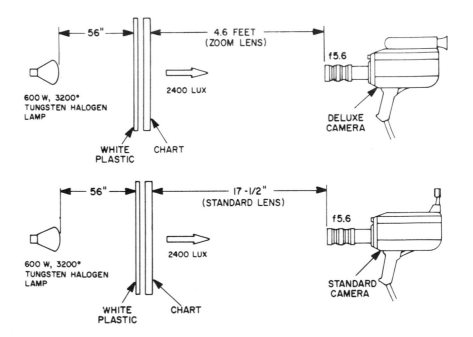

Fig. 10-11 Illumination test setup (*Courtesy,* Zenith Radio Corp.)

The camera should be securely mounted on a tripod and positioned so that the field of view just includes the chart area. The camera lens aperture must be set to f:5.6. The camera lens focus should be set to the distance between the camera lens and the chart. If the standard lens (not zoom lens) is used, this minimum distance is about 17½ in. For the zoom lens camera, it will be about 4.6 ft.

Overall Adjustment Procedure

The complete adjustment procedure must be performed in the correct order to ensure proper adjustment. Since many adjustments interact, the same adjustment may be repeated several times throughout the procedure. Before making any adjustments, be sure that all test equipment is accurate and operating properly.

The following adjustment procedures are to be performed in sequence:

1. Remove the camera lens, case, and insulator board.
2. Remove screws to allow access to both sides of the PCB containing the sync encoder.
3. Reattach the lens to the camera and set aperture to C (closed iris).
4. Connect equipment as shown in Fig. 10-12.
5. Set up light source as specified under illumination setut up in Fig. 10-11 using the gray scale.
6. Turn on power adapter and allow camera to warm up for 10 min.
7. Adjustment of +9 V: Measure the dc voltage at pin 8 of IC201, shown on the regulator PCB in Fig. 10-9. Remove the viewfinder support bracket and the regular electrostatic shield to gain access to VR401, and adjust it to obtain +9 V at pin 8 of IC201.
8. Target voltage: Turn VR107 (ASC control) on video PCB in Fig. 10-9 fully clockwise as seen from the foil side of the PCB in order to disable the automatic sensitivity control (ASC) and apply maximum voltage to the target electrodes.
9. Beam adjustment: With lens iris closed, turn the 600-W tungsten halogen lamp On. Make certain that the illumination setup is as described in Fig. 10-11. Use the gray scale chart, and set Brightness control VR109 on the video PCB to normal and Color Temp control VR104 to the third mark from a fully counterclockwise position. Turn VR102 (beam control on sync-encoder PCB) fully clockwise as seen from the foil side of the PCB. Set the lens iris to the point at which video output is just below overloading and adjust VR102 to the point at which video output just begins to drop.
10. Magnetic focus: Use the resolution chart for this adjustment. Adjust focus control VR101, located on the regulator PCB, and the lens focus for the best focus. Adjust the alignment magnet rings, located at the rear of the yoke assembly of the vidicon, so that the center of the picture does not move when VR101 is turned. Only the edge of the picture rotates, as shown in Fig. 10-13.
11. Subcarrier frequency check: Measure the frequency at pin 9 of

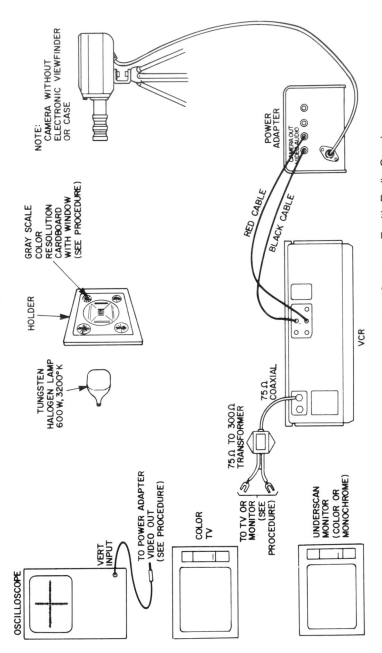

Fig. 10-12 Camera adjustment test setup (*Courtesy, Zenith Radio Corp.*)

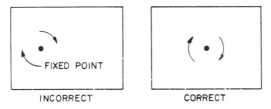

Fig. 10-13 Magnetic focus adjustment (*Courtesy,* Zenith Radio Corp.)

IC201 on the sync-encoder PCB. It should measure 3.579545 MHz. If the frequency is off by more than ±50 Hz, adjust VC101, located on the sync-encoder PCB.

12. Burst and sync-level check: Disconect the power-adapter video output cable and connect the oscilloscope, terminated into 75 ohms, to power the adapter video output connector. Measure the burst level and sync level for a 0.3 + 0.1 − 0.05 Vpp level. The sync level and the burst level are controlled by resistors associated with Q208 and Q209, respectively, on the sync-encoder PCB. They are factory selected to produce the proper output levels.

13. Beam adjustment: Repeat procedure 9 above.

14. Aperture indicator: Use the gray scale chart and adjust the lens iris so that video output drops to 0.35 Vpp (not including the sync level). Adjust VR106 on the video PCB so that the aperture indication LED just begins to blink on and off.

15. + 9-V check: Repeat procedure 7 above.

16. Subcarrier frequency check: Repeat procedure 11 above.

17. Burst and sync level check: Repeat procedure 12 above.

18. Beam adjustment: Repeat procedure 9 above.

19. Magnetic focus adjustment: Repeat procedure 10 above.

20. Back focus adjustment (*Procedure for camera with electronic viewfinder*):Use resolution chart. Remove the rubber lens hood and install the 58-mm screw-in + 2 close-up lens. Position the camera so that the distance between the center of the close-up lens and the test chart is ½ m (19.7 in.). Set the lens focus to infinity. Adjust the lens iris to f:2.0. Use the telephoto position of the zoom, and adjust VR107 on the video PCB to produce a 1.0-Vpp video signal output (sync tip to peak white). Loosen the back focus lock screw, and adjust the back focus adjustment screw for best focus. Reposition the camera 5 ft away from the test chart, and remove the 58-mm screw-in + 2 close-up lens. With the lens iris set to f:2, focus the camera on the test chart, and check the reading on the focus ring. It should read 5 ft. If it does not, repeat the procedure. When adjustment is complete, turn VR107 fully clockwise as seen from the PCB foil side. Set the iris to F:5.6.

21. Beam adjustment: Repeat procedure 9.

22. Picture tilt: Loosen the yoke-locking screw, and rotate the yoke to the point at which the base line of the gray scale chart is perfectly horizontal.

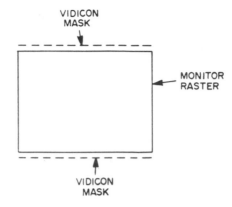

Fig. 10-14 Vertical size adjustment (*Courtesy,* Zenith Radio Corp.)

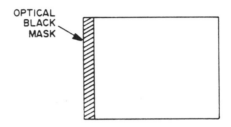

Fig. 10-15 Horizontal size adjustment (*Courtesy,* Zenith Radio Corp.)

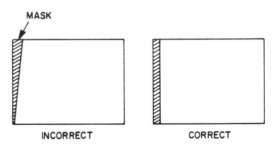

Fig. 10-16 Optical black mask positioning (*Courtesy,* Zenith Radio Corp.)

23. (a) Vertical scanning: Use an underscanned monitor and resolution chart. Set the iris to f:5.6. Adjust VR201 (vertical linearity), VR202 (height), and VR203 (vertical center)—all located on the regulator PCB—so that the large circle in the center of the chart is reproduced as a circle. VR202 and VR203 should be adjusted so that the vidicon mask is just beyond visibility. See Fig. 10-14.

(b) Horizontal scanning: Use an underscanned monitor and resolution chart. Set iris to f:5.6. Adjust VR101 (horizontal centering), VL101 (horizontal linearity), VL102 (width)—all located on the sync-encoder PCB—so that the large circle is reproduced as a circle and the vidicon mask on the right side is just beyond visibility while the optical black mask on the left side is slightly

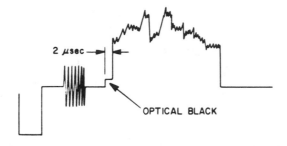

Fig. 10-17 Optical black pedestal (*Courtesy,* Zenith Radio Corp.)

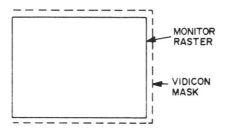

Fig. 10-18 Vertical positioning (*Courtesy,* Zenith Radio Corp.)

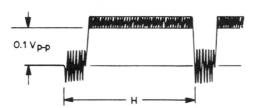

Fig. 10-19 Dark current adjustment waveform (*Courtesy,* Zenith Radio Corp.)

visible (less than one-twentieth of the picture area). See Fig. 10-15.

24. (a) Horizontal position of black mask: Use an underscanned monitor. Turn VR101 (horizontal centering located on sync-encoder board) to a point at which the optical black mask can be seen on the left side of the monitor (see Fig. 10-16). Loosen vidicon clamp screw, and rotate vidicon until black mask is parallel with the left side of the monitor. Observe one horizontal line by using an oscilloscope, and adjust VR101 so that the optical black (black mask) produces a pedestal of from 1.5 to 2 µsec. See Fig. 10-17. Then adjust VL101 (widths) so that the vidicon mask cannot be seen on the right side of the monitor.

(b) Vertical position of black mask: Adjust VR202 (height) and VR203 (vertical centering) so that the vidicon mask on top and bottom is just beyond visibility. See Fig. 10-18.

25. Back focus: Repeat procedure 20.

26. Check video output: Look at monitor and determine if picture linearity and size are acceptable.

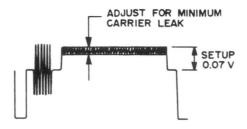

Fig. 10-20 Black balance waveform (*Courtesy,* Zenith Radio Corp.)

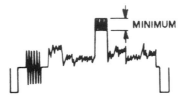

Fig. 10-21 Chroma balance waveform (*Courtesy,* Zenith Radio Corp.)

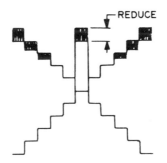

Fig. 10-22 White balance waveform (*Courtesy,* Zenith Radio Corp.)

27. Vidicon spots: Look at the monitor and check to see if there are any unusual spots or burns in the vidicon. Use the resolution chart.
28. Set color temperature: Rotate VR104 (Color Temp control) to the third mark from a fully counterclockwise position and set VR109 Brightness to center position.
29. Dark current adjustment: Close the iris (set to C) and observe the waveform at the emitter of Q104 with an oscilloscope. Adjust VR101 (dark current adjustment) so that black level with respect to sync is 0.1 Vpp. See Fig. 10-19.
30. Black balance: Repeat procedure 28 and then set Brightness control (VR109) to center position. Adjust VR103 (red setup), VR108 (blue setup), and VR112 (green setup) for minimum carrier leak by looking at the waveform with an oscilloscope that is connected to the video output terminals. See Fig. 10-20.
31. Chroma balance: Use the window chart and observe the video output with an oscilloscope. Look at the window portion of the

waveform, and adjust VR113 (R–Y balance) and VR114 (B–Y balance) on the video PCB for minimum carrier leak on the white portion of the window. See Fig. 10-21.

32. White balance: Use the gray scale chart and observe the video output with an oscilloscope. Adjust VR102 (red gain) and VR105 (blue gain) on the video PCB for minimum carrier leak on the upper chart of the gray scale chart. See Fig. 10-22.

33. Cross talk balance: Use the window chart and adjust VR110 (red impedance) and VR111 (blue impedance) on the video PCB minimum color smear (while observing color monitor) on the edges of the window.

34. ASC (automatic sensitivity control): Use the gray scale chart with standard lighting conditions and the iris set to f:5.6. Adjust VR107 (ASC control for 0.7-Vpp video output level.

35. Black balance adjustment: Repeat Procedure 30.

36. White balance adjustment: Repeat Procedure 32.

37. Aperture indicator adjustment: Repeat Procedure 14.

38. Color temperature check: Check to see if picture turns orange by turning on the Color Temp control fully clockwise.

39. Setup check: With iris closed, turn Brightness control fully counter-clockwise, and measure the black level setup with an oscilloscope looking at the video output, which should measure 50 mV. See Fig. 10-20. Turn Brightness control fully clockwise and measure again. This time it should read 90 mV. Set Brightness control to center position and measure again. The reading should now be above 70 mV.

40. Check beam level adjustment: Repeat Procedure 9.

41. Check video output: Use color bar chart and make sure that colors are reproduced properly.

Adjustment Procedures for Electronic Viewfinders

Warning: Dangerously high voltages exist in electronic viewfinders. Use extreme care when servicing.

Equipment Required

Video camera
Tripod
600-W tungsten-halogen lamp (3200°K)
White plastic diffuser (Kolonite plexiglas, #W-2447, P-80 finish)
Chart holding frame (any holder or clamp which will hold the diffuser and chart)
Resolution test pattern (EIAJ Type A)
Monitor with video inputs or VCR and black-and-white or color TV receiver
Video cassette recorder (not needed if resolution chart is used)
Pattern generator (not needed if resolution chart is used)

Procedure

Camera and monitor must be properly adjusted before starting the following procedures for adjusting the electronic viewinder. These procedures must be performed in the sequence given:

1. Remove the electronic viewfinder cover and remount the viewfinder on the camera.
2. Connect the equipment as shown in Fig. 10-23, using either the tungsten-halogen lamp and resolution chart or a pattern generator connected through a video cassette recorder.
3. Mount the resolution test pattern and diffuser on the chart holding frame.
4. Set the camera 10 ft from the chart holding frame.
5. Adjust the camera focus or proper distance and set the aperture to f:5.6.
6. Observe the monitor for correct picture.
7. Observe the viewfinder picture and adjust its horizontal hold control for horizontal synchronization—that is, elimination of left or right rolling.
8. Adjust the vertical hold control of the viewfinder for vertical synchronization (elimination of vertical roll).
9. Adjust the vertical height control of the viewfinder so that the

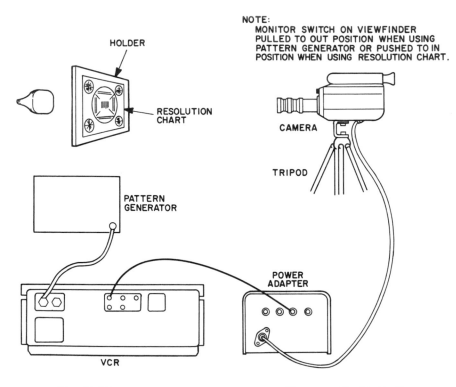

NOTE:
MONITOR SWITCH ON VIEWFINDER
PULLED TO OUT POSITION WHEN USING
PATTERN GENERATOR OR PUSHED TO IN
POSITION WHEN USING RESOLUTION CHART.

HOLDER

RESOLUTION CHART

CAMERA

TRIPOD

PATTERN GENERATOR

POWER ADAPTER

VCR

Fig. 10-23 Electronic viewfinder test setup (*Courtesy*, Zenith Radio Corp.)

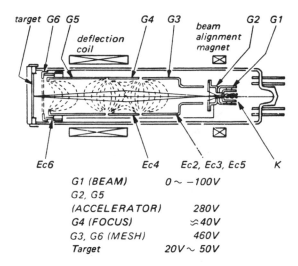

G1 (BEAM)		$0 \sim -100V$
G2, G5		
(ACCELERATOR)		280V
G4 (FOCUS)		$\approx 40V$
G3, G6 (MESH)		460V
Target		$20V \sim 50V$

Fig. 10-24 Monochrome vidicon (*Courtesy*, Zenith Radio Corp.)

vertical height is approximately 110 percent, and adjust the vertical linearity control to obtain a circular test pattern.

10. Adjust the focus control of the viewfinder for optimum picture clarity.

11. Adjust the brightness control of the viewfinder for optimum brightness.

12. Adjust the contrast control for optimum picture density.

13. If the raster is inclined or off center, loosen the deflection coil screw and position the yoke on the CRT to correct the inclination. Also, adjust the centering magnets to provide horizontal centering.

Monochrome TV Cameras

Monochrome TV cameras suitable for home or personal portable use employ vidicon tubes whose target construction is much simpler than that of the single tubes used in color TV cameras. They may be either of the magnetic focus or electrostatic focus type. The latter is preferred because its power consumption is lower, requiring neither the focus coil nor its accompanying circuitry. In this design, the electron beam emitted by cathode K is accelerated by electrode G2 after passing control grid G1, as shown in Fig. 10-24. Then it passes through a 30-micron beam-limiting aperture to form a small-diameter electron beam, the latter being focussed by the electrostatic lens consisting of electrodes G3, G4, and G5. Electrodes G5 and G6 form an independent collimating lens that ensures that the input angle of the electron beam against the target is 90 deg at any point on the target. High resolution at the corners of the picture is obtained in this way.

The magnetic deflection coil is seen between the focusing electrodes and collimating electron lens. A beam-alignment magnet is also seen at the right in Fig. 10-24. Its function is to position the electron beam in the center of the vidicon diameter. Incorrect adjustment of the beam-align-

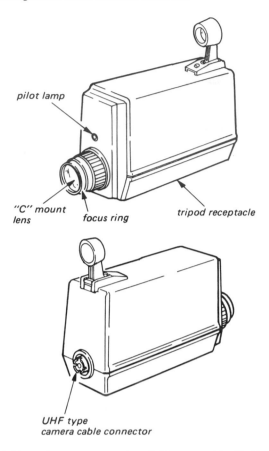

Fig. 10-25 Monochrome camera head (*Courtesy,* Zenith Radio Corp.)

ment magnet results in extremely poor resolution. Proper maintenance of the camera requires careful adjustment of this magnet.

In the Zenith example that we are using here, the camera pickup tube is a ⅔-in. vidicon with electrostatic focus and magnetic deflection as described above. It has a Sb2S3 photoconductive layer and a 1000-lines/inch copper target mesh.

The Camera Head

As shown in Fig. 10-25, the camera head is fitted with a "C" mount lens. Wide-angle, telephoto, and zoom lenses are available as optional accessories. As with most monochrome cameras, the viewfinder is optical. A separate power unit, shown in Fig. 10-26, is furnished.

The camera head houses the vidicon, its deflection yoke, and various sections of circuits on separate PC boards. In the case of Zenith's model JC-500, three boards hold the video signal circuits, the deflection circuits, and the high-voltage power supply circuits connected to each electrode of the vidicon. Referring to Fig. 10-27, it is seen that the video signal circuits consist of video amplifiers, peak clipper, target AGC, video output, clamp, sync mixer, and ac-dc converter. The deflection circuits include the 31.5-

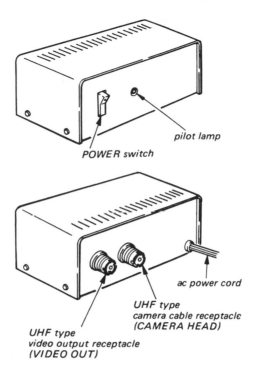

pilot lamp

POWER switch

ac power cord

UHF type
camera cable receptacle
(CAMERA HEAD)

UHF type
video output receptacle
(VIDEO OUT)

Fig. 10-26 Power unit for monochrome camera (*Courtesy,* Zenith Radio Corp.)

kHz master oscillator, f/2 counter, f/525 counter, horizontal deflection, vertical deflection, blanking amplifier, and the 12-volt regulator with ripple detector and cancel circuits. A light-emitting diode (LED) is included as a pilot lamp.

Maintenance of the Monochrome Camera

The following precautions should be observed to prevent damage to the vidicon:

1. Do not jar the camera. Avoid handling the camera with the vidicon faceplate pointing downward.
2. Before applying power to a defective camera of unknown condition, turn the Beam control fully clockwise. The vidicon may be damaged if power is applied in the absence of deflection current.

Vidicons should be replaced only in accordance with the detailed instructions given in the manufacturer's service manuals.

Vidicon Adjustments

Deflection checks: The main point about these checks is to make certain that the deflection circuits of the camera are working properly. If the condition of deflection is unknown, check the deflection waveforms. If normal waveforms are observed, turn off the camera and push the socket onto the vidicon.

Vidicon high-voltage check: Check the dc voltage at each vidicon electrode for proper vidicon drive.

Test Jig

To view a test pattern, set up the camera in a test jig as shown in Fig. 10-28.

Optical Focus

There are two possible procedures for optical focus: (1) if a zoom lens is not available, and (2) if it is. If a zoom lens is not available, take the following steps:

1. Turn on the camera and point it at a distant object (farther than 50 ft away).
2. Set the standard lens to infinity and adjust the Focus control for sharpest focus. It is important to have electrical focus set correctly before attempting the remaining adjustments.
3. Loosen the screw on the lens mount ring and turn the lens mount ring for sharpest focus.
4. Point the camera at a test card 2½ ft from the lens. Bring the card into focus by adjusting the focus ring. Reset the lens mount if necessary.

Tilted Picture

If the picture as viewed on the monitor screen is tilted, the deflection yoke must be rotated about its axis. Loosen the screws that secure the yoke, rotate the yoke until the picture is upright, and tighten the screws.

If a zoom lens is available, take the following steps:

1. Remove the standard lens and install the zoom lens.
2. Place the camera in the test jig shown in Fig. 10-28, or mount the camera on a tripod and point it at a test pattern 22 ft from the lens.
3. Connect a monitor to display the video output.
4. Set the lens zoom control to the telephoto position and adjust the lens focus control for the sharpest picture.
5. Set the lens zoom control to the wide-angle position. Loosen the screw in the lens mount ring and adjust the ring for optimum focus.
6. Set the lens zoom control to the telephoto position again and check the focus. The line should track throughout the entire zoom range. If focus does not track correctly, repeat steps 4 and 5 two or three times.
7. When proper tracking has been established, tighten the screws on the lens mount ring. Check to make sure that the picture on the monitor screen is not tilted. If it is, adjust the tilt according to the procedure outlined above.

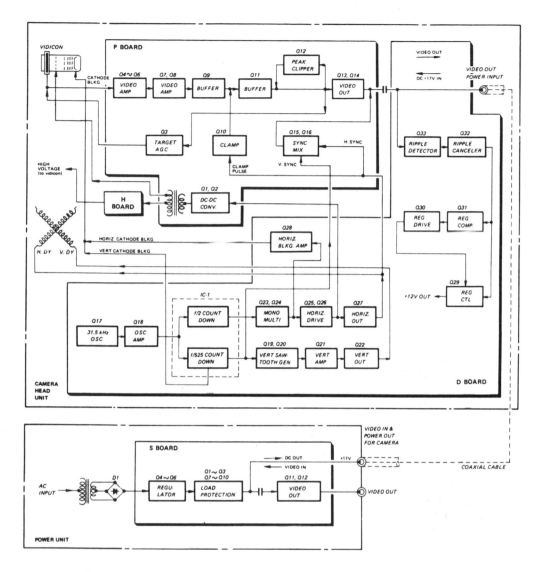

Fig. 10-27 Block diagram of black-and-white camera (*Courtesy,* Zenith Radio Corp.)

Video Adjustments

Target voltage, focus, beam control, and sync level determine the character of the video output signal. Correct target voltage settings are important because they determine overall sensitivity, video output level, and image persistence. The need for proper focusing is obvious. Excessive beam current produces poor resolution. Proper sync level is essential to ensure a stable picture.

After the preliminary setup, video adjustments should be made in the following order: (1) sync level adjustment, (2) beam control adjustment, (3) target voltage adjustment, and (4) electrical focus adjustment. Voltage levels, test points, and other details for these adjustments must be

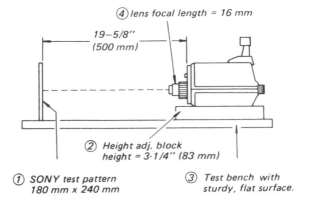

④ *lens focal length = 16 mm*

19–5/8''
(500 mm)

② *Height adj. block*
    height = 3-1/4'' (83 mm)

① *SONY test pattern*   ③ *Test bench with*
    180 mm x 240 mm   *sturdy, flat surface.*

Fig. 10-28 Test jig setup for black-and-white camera (*Courtesy*, Zenith Radio Corp.)

obtained from the manufacturer's service manual for the specific camera to be tested. Electrical focus adjustment is described here because it does not require such specific information from a service manual.

Electrical Focus Adjustment

1. Place the camera in a test jig as shown in Fig. 10-28, or mount the camera on a tripod and point it at the test pattern. Adjust the distance between the lens and the test pattern so that the test pattern fills the monitor screen.
2. Provide illumination of more than 1000 lux. Warm up the camera for more than 5 min.
3. Adjust the optical focus control for best focus.
4. Adjust the electrical focus control for the sharpest picture on the monitor.
5. Repeat steps 3 and 4 several times, if necessary.
6. Rotate the vidicon alignment magnet for optimum resolution at the corners of the picture.

Deflection Adjustments

Sync generator frequency, horizontal centering, horizontal size, vertical centering, vertical size, and vertical linearity determine the vidicon deflection rate and size and the linearity and position of the picture. The sync generator frequency should be adjusted in accordance with the details given in the manufacturer's service manual. Horizontal and vertical adjustments are usually related to controls that are available on any TV camera. These adjustments will now be described.

Horizontal Centering Adjustment

1. Set up the camera to display the video output on a monitor.
2. Adjust the horizontal size control on the deflection PCB until the black target ring appears at both sides of the monitor screen, as shown in Fig. 10-29.

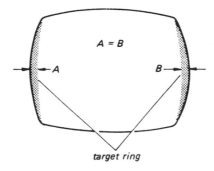

Fig. 10-29 Horizontal centering adjustment (*Courtesy*, Zenith Radio Corp.)

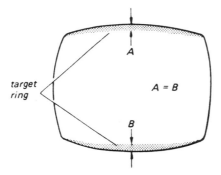

Fig. 10-30 Vertical centering adjustment (*Courtesy*, Zenith Radio Corp.)

3. Turn the horizontal centering control on the deflection PCB so that the black target ring appears an equal amount at the right and left sides of the monitor. If the black target ring is not visible on the monitor screen, turn the horizontal centering control until it appears at the left side. Then turn the control counterclockwise until it appears at the right side of the monitor screen. Set the control midway between the settings at which the target ring appears.

Horizontal Size Adjustment

1. Place the camera in the test jig (Fig. 10-28), or point the camera at a test pattern located 18⅝ in. from the lens.
2. Connect a video monitor to the display video output.
3. Adjust the horizontal size control on the deflection PCB so that the width of the test pattern fills the screen horizontally.

Vertical Centering, Size, and Linearity Adjustments

1. Set up the camera to display the video output on a monitor. Confirm that the monitor has good linearity.
2. Place the camera in the test jig (Fig. 10-28) or point the camera at a test pattern 18⅝ in. away from the lens.

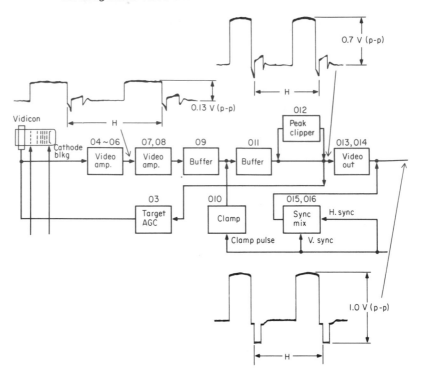

Fig. 10-31 Key waveforms important to troubleshooting

3. Adjust the vertical size control on the deflection PCB until the black target ring appears at the top and bottom of the monitor as shown in Fig. 10-30.
4. Adjust the vertical centering control on the deflection board so that the black target ring appears by an equal amount at the top and bottom of the screen.
5. Adjust the vertical size control so that the height of the test pattern fills the screen vertically.
6. Adjust the vertical linearity control for the best linearity on the display. If adjustment of this control affects vertical size, repeat steps 5 and 6 as required.

Power Supply Adjustments

Cameras have one or more regulated power supplies, which should be checked to determine whether all voltages correspond to the correct ones given in the service manual of the TV camera manufacturer or supplier.

Video Checks

The following tests will help to localize trouble so that one can service monochrome cameras more efficiently:

1. Set up the camera with a monitor.

2. Turn on the camera and check the video output. If the output signal contains sync and blanking but no video, the trouble is confined to the video circuits or the vidicon itself.

3. To localize loss of video in the vidicon or the video amplifier, place your finger near the target lead at the front of the vidicon deflection coil. If the video circuits are working normally, a herringbone interference pattern will appear on the monitor screen, confining the trouble to the vidicon or the vidicon drive high-voltage supply. If the pattern does not appear, the trouble lies in the video amplifier section.

4. If the trouble lies in the vidicon, make sure that the vidicon has the necessary operating voltage. Target voltage should not be measured at the target ring, because the voltmeter will produce loading at that point and provide an erroneous reading. Target voltage should be checked at its source in the regulated power supply.

Key waveforms at certain points in the video amplifier are shown in Fig. 10-31. They are important for signal tracing through these circuits. Here they are related to circuit blocks. The exact test points can be found by referring to the service manual of the specific camera to be serviced.

Appendix A

*The Automatic Assembly Recording System**

In using a portable video recorder in combination with a video camera, one often makes short intermittent recordings. Ordinarily, the camera's pause switch is used concurrently. When such recorded tapes are played back, very annoying noises and flickers will occur in the pictures at the boundary between successive recording shots. This effect is due to the overlapping of vertical sync signals at the boundary of two recorded shots, which may be explained by referring to Fig. A-1.

When a recording is made while using the Pause button provided on the video camera, the operation mode of the VTR changes from Record to Pause and again into Record as shown in Fig. A-1. In the Pause mode, the

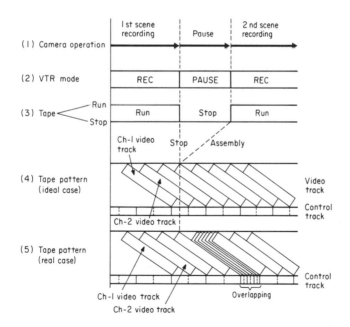

Fig. A-1 Timing and tape patterns of intermittent recording

*Appendix A is excerpted from "Automatic Assembly Recording System Helical Scan Type VTR" by I. Fukushima, H. Nishijima, & H. Yokota, *Transactions on Consumer Electronics*, vol. CE-25, © 1979, IEEE, and is reprinted with the permission of Hitachi, Ltd.

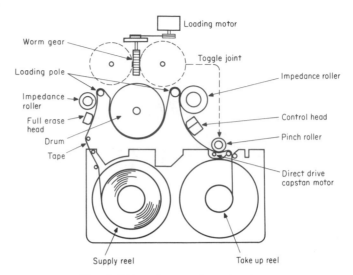

Fig. A-2 Tape transport mechanism

loading motor slightly reverses so that the pinch roller is released from the capstan shaft to suspend tape transport. When the Pause button is released, the loading motor turns in the normal direction so that the pinch roller is again pressed against the capstan shaft to restart the tape transport. In such an intermittent recording, no picture disturbance will take place at each boundary between two recorded shots if a proper recording is made with the Channel-1 and Channel-2 heads recording alternately as shown in Fig. A-1(4). However, in most conventional VTRs, the recording patterns of the vertical sync signal on the control track partially overlap each other at the boundary of two recording shots and so do the video signal patterns, as shown in Fig. A-1(5).

It is apparent that the noises and flickers present in the playback pictures cannot be due to the absence of video signal at the boundary but to the disturbed recording track configuration at such boundaries, which the two rotary heads cannot retrace precisely in the playback mode. For instance, if the Channel-1 head retraces the Channel-2 video track and Channel-2 head retraces the Channel-1 video track at a recording shot boundary, the playback output will be considerably lowered because of the azimuth recording system. As a result, the playback picture is subject to white noises or flickers because of the lowered vertical sync signal level. This phenomenon continues for about 0.2 to 2 sec until the servo system synchronizes with the vertical sync signal on the second shot.

Conventional Assembly Recording Systems

Although assembly recording is not used in conventional home video recorders, it is common to broadcast and industrial recorders and can be classified in two categories related to their operating principles, as follows:

1. When the Pause button is depressed, a short amount of tape is rewound at the same time. When the Pause button is released, the

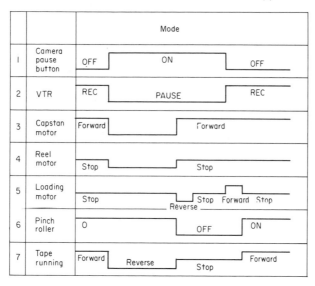

Fig. A-3 Motor operation and tape running mode for intermittent recording

recorded vertical sync signal on the control track on the first shot and that of the video signal on the second shot are synchronized during the period in which the previously rewound tape runs forward before recording is started. This system is commonly used for ¾-in. VTRs.

2. When the Pause button is depressed, the tape stops immediately after the recording by the specific channel of recording heads is completed. When the Pause button is released, the tape starts and the recording is begun immediately after the other channel head comes precisely to the head of the video track. The relative positional relationship between the video tracks and the video heads is thus maintained constant.

In the first system above, the synchronization control to synchronize the vertical sync signal recorded on the control track for the first shot with that of the video signal of the second shot includes the capstan. Since the capstan has a large amount of inertia, it takes approximately 0.5 to 2.0 sec to bring the system into complete synchronization. Consequently, it is necessary to rewind the tape by an amount corresponding to this synchronization time. As a result, the video signal for the second shot cannot be recorded during this time. This is particularly disadvantageous if paused recording is used frequently with a video camera.

The second synchronization system mentioned above is the better of the two in that no picture disturbance takes place. However, in order to stop or start the tape instantaneously a solenoid is required for pinch roller control. In addition, the system also requires a brake solenoid to hold the video track on the tape always in the correct positional relationship with respect to the video heads.

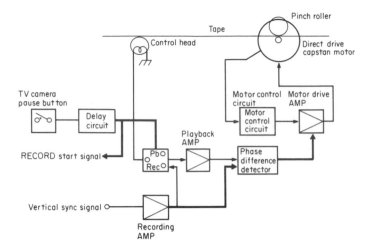

Fig. A-4 Capstan drive system for assembly recording

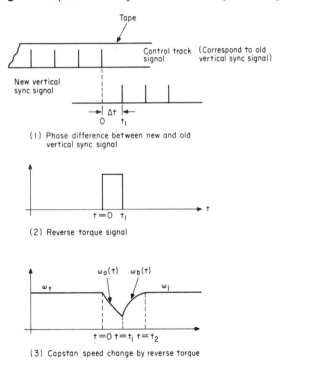

(1) Phase difference between new and old
    vertical sync signal

(2) Reverse torque signal

(3) Capstan speed change by reverse torque

Fig. A-5 Capstan speed change by reverse torque proportional to the phase difference

Hitachi Assembly Recording System for Portable Recorder

In the tape transport system shown in Fig. A-2, when the Pause button is depressed, the capstan motor reverses its turning direction to rewind a small amount of tape, and, at the same time, the reel motor takes up this rewound tape on the supply reel. Then the loading motor turns slightly in the reverse direction so that the pinch roller is released from the capstan to stop the tape transport. The capstan motor then turns in the normal direction. When the Pause button is released, the loading motor rotates in the normal direction so that the pinch roller is pressed against the capstan shaft to start the tape transport. A vertical sync signal is now played back from the tape previously rewound and is instantaneously synchronized with the vertical sync signal for the coming shot before recording is started. The timing sequence for the above pause operation is shown in Fig. A-3.

In order to obtain instantaneous synchronization between two vertical sync signals for two different recording shots, the system shown in Fig. A-4 was arranged. The shaded portions in the figure represent additions to the conventional system. In this new system, the tape is run, but the recording of both the audio and video signals is suspended. During this period, the vertical sync signal on the control track shown in Fig. A-5 (1) is played back by the control head and amplified by the playback amplifier. The first output pulse from this playback amplifier sets the phase detector, which consists of RS flip-flops. Then the phase detector is reset by the first pulse of the vertical sync signal separated from the video signal for the second shot so as to obtain the maximum reverse torque, which is proportional to the phase difference between the two vertical sync signals [see Fig. A-5(2)]. When this signal is applied to the direct-drive capstan motor, the motor speed changes, as shown in Fig. A-5(3).

Tape speed is thus momentarily reduced by the reverse torque signal applied to the motor so as to synchronize the vertical sync signal for the first shot to be recorded with that for the second shot. The Pause button signal is delayed for the period required for this synchronization process before the recording of the second shot is started. The key point of this system is to reduce the tape speed within this period to correspond to the phase difference between two vertical sync signals for two different recording shots so that the two vertical sync signals will be synchronized within a phase difference of ± 90 deg. Actually, the time corresponding to the maximum phase difference of 360 deg is 33.3 ms at the most. One design question was whether the rotation of a capstan with a large moment of inertia could be reduced within a period less than 33.3 ms. Use of a quick-response direct-drive motor to drive the capstan achieved a sufficient reduction.

The development of this portable recorder confirmed that the noises and flickers in playback pictures at each boundary of two successive recording shots become almost invisible if the two vertical sync signals for the two successive shots are phased within ± 90 deg at such boundaries. In its final form, this portable recorder incorporates an automatic assembly recording system, which works in the following way:

1. When the Pause button on the video camera is depressed, the capstan motor is reversed at a constant speed in order to rewind a specific amount of tape onto the supply reel. After the rewinding is completed, the tape is stopped, and then the capstan motor is restored to forward rotation.

2. When the Pause button is released, the tape is first run in the playback mode, and a reverse torque is applied to the capstan motor for a period that corresponds to the phase difference between the already recorded vertical sync signal for the previous shot and that for the next shot to be recorded.

3. After the phase difference between the two vertical sync signals has been controlled within ± 90 deg, the VTR is put into the record mode to start recording.

*Appendix A is excerpted from "Automatic Recording System Helical Scan Type VTR" by I. Fukushima, H. Nishijima, & H. Yokota, *Transactions on Consumer Electronics*, vol. CE-25, © 1979, IEEE, and is reprinted with the permission of Hitachi, Ltd.

Appendix B
Special Considerations for Cable TV

When a video recorder and receiver are connected to an outside or indoor antenna for normal reception of broadcast programs, it is possible to record one channel and view another at the same time. Also, with a programmable video recorder, a number of channels can be selected as desired. However, neither of the above is possible when a VCR and TV receiver are connected to a typical cable TV outlet unless some special arrangements are made because channel selection in cable TV systems is often accomplished by tuning a converter located between the cable outlet and the VCR and TV receiver. This converter usually converts all incoming channels to a single channel at its output (often channel 3). The incoming channels, which can only be selected at the converter, can be received or recorded only by tuning the TV receiver or the VCR to a single-output channel, such as channel 3.

A way around the above limitation in cable recording has been found by substituting an up-converter for the conventional cable TV converter. This rather small and inexpensive electronic package simply converts all VHF TV signals (special cable TV channels as well as the regular VHF channels) to frequencies that are tunable in the UHF band range of the VCR receiver. Hence, as shown in Table B1, a set of

Table B1

VHF channels			Midband channels			Superband channels		
VHF channel	Up-converted UHF channel		Channel	Up-converted UHF channel		Channel	Up-converted UHF channel	
	No. 1	No. 2		No. 1	No. 2		No. 1	No. 2
2	36	38	A	47	49	J	63	65
3	37	39	B	48	50	K	64	66
4	38	40	C	49	51	L	65	67
5	40	42	D	50	52	M	66	68
6	41	43	E	51	53	N	67	69
7	56	58	F	52	54	O	68	70
8	57	59	G	53	55	P	69	71
9	58	60	H	54	56	Q	70	72
10	59	61	I	55	57	R	71	73
11	60	62				S	—	74
12	61	63						
13	62	64						

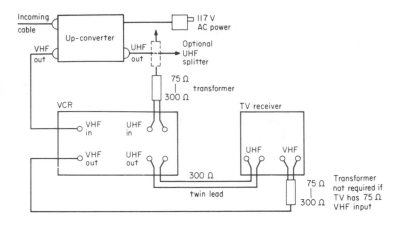

Fig. B-1 Connection of Philips channel plus up-converter from TV cable to VCR and TV receiver

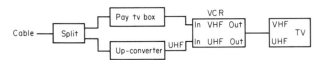

Fig. B-2 Connection of up-converter when pay TV is involved (*Courtesy,* ETCO Electronics)

UHF channels on the VCR tuner will correspond to certain VHF broadcast channels as well as the special cable TV channels. There are several brands of up-converters on the market, and while there is a considerable similarity in their designs, there are at least two different sequences of UHF channel settings for given VHF, midband, and superband channels in the signals from such converters.

A Philips Channel Plus Up-Converter connected as shown in Fig. B-1 will permit the following operation:

1. The TV receiver may receive all channels on its UHF tuner with the VCR antenna switch in either the VCR or TV receiver positions. Thus, one channel may be viewed while another is being recorded.
2. The TV receiver may receive VHF channels or VCR output (that is, channel 3 or 4) by using its VHF tuner, depending upon the VCR antenna switch position.
3. The VCR may receive all channels on its UHF tuner and thus provide full use of VCR programmable features. The No. 2 UHF channel conversion shown in Table B1 is provided. In this case, the up-converter covers a VHF range of 50–280 MHz and a UHF range of 610–840 MHz.
4. The VCR may also receive VHF channels on its VHF tuner.
5. Some channels will appear in more than one position on the VHF and UHF tuners. The position with the best picture quality should be used.

Other makes of up-converters may require different connections, which should be determined from the hookup instructions furnished with them. Also, the specific UHF channel conversion may vary.

The up-conversion process may result in interference and herringbone patterns on some channels, especially when multichannel cable systems are congested. Isolation and separate shielding in the internal design of the up-converter will minimize such effects. A further step to counter such effects is to keep all RF cables between units as short as possible and away from each other as much as possible to avoid external coupling.

When pay TV is involved, the VCR and TV receiver may be connected as shown in Fig. B-2. Connections shown here are for the VIDCOR 2000 (ETCO VA950) Up-Converter. In this case, the pay TV is tuned on the channel output of its converter (usually 3, 4, or 12). All other channels are tuned on UHF. Switching between the VCR and the TV receiver should be done in accordance with the VCR manufacturer's instructions.

Glossary

ACC (Automatic Color Control)   Circuitry used to maintain an overall constant color signal level in the color circuits. Since there are at least two video heads in modern VCRs, differences in their playback characteristics can introduce field-to-field variations in the playback signals. By maintaining equal color levels from field to field, the ACC eliminates the 30-Hz flicker that may occur because of variations between the video heads.

ACK (Automatic Color Killer)   Circuitry to shut down color processing during the playback of monochrome signals. In addition to eliminating spurious effects from color channels, this circuitry sometimes increases the FM bandwidth for monochrome signals to provide a black-and-white picture of higher definition.

Adjacent track   The video track to the immediate right or left of the track being processed.

AFC (Automatic Frequency Control)   Circuitry used to phase-lock the color circuits to either the record or playback color signal in order to achieve a stable color signal. Increased stability of the TV horizontal oscillator is achieved by triggering this oscillator circuit by means of the sync signal (at 15.75 kHz) separated from the composite video signal. However, if only the horizontal sync signal separated from the composite video signal is used to trigger the oscillator circuit directly, noise components within the sync signal's frequency band will cause nonuniformity in the oscillator cycle, with a corresponding degradation in picture stability. The AFC detects phase differences between the sync signal and oscillator frequency and automatically phase locks the oscillator.

AFT (Automatic Fine Tuning)   A special circuit found in the receiver portion of home VCRs, which makes the local oscillator of the TV tuner follow the selected channel in order to produce a stable IF frequency.

AGC (Automatic Gain Control)   Circuitry used to maintain an overall constant picture level in the luminance circuits by minimizing the effects of input level fluctuations.

APC (Automatic Phase Control)   Circuitry that refines the action as described under AGC above to remove most of the residual frequency and phase errors in the playback signal.

Azimuth   A term used to describe the angular tilt of the gap of the recording head when it is viewed straight on. This tilt may be to the right or the left.

Azimuth recording   A form of recording found in most home VCRs in the longer play modes. It makes overlapping of adjacent tracks possible by taking advantage of the azimuth loss, which occurs particularly at higher frequencies when the azimuh angle of the recording head gap is different from that of the playback head gap. By giving video heads A and B a relative angle difference (14 deg for Betamax and 12 deg for VHS) when the track recorded by head A is played back, the track recorded by head B will not be reproduced because of azimuth loss. In the same way, during playback of track B, the signal of track A will not be picked up. Thus, extended playback and recording times become possible since empty space between tracks is not needed to prevent crosstalk.

Beta format   An inclusive term describing the home VCR system developed by Sony.
Blanking   TV scanning goes from left to right and from top to bottom. When the scanning beam gets all the way to the right (or bottom), it must go back to the left (or top). This retrace time is called the *blanking period*. To maintain good picture quality, the electron beam is shut off during the instant it takes to get to the beginning of the next scanning line. The blanking period may be used to eliminate undesirable picture effects by moving them into the blanked portion if possible.
Burst   *See* Color Burst

Capstan   A small rotating metal dowel that drives the recording tape to assure positive tape movement. The video heads travel past the tape at a much higher speed than the capstan drives the tape. The high-speed rotating drum carries these heads at a speed that will ensure a sufficient video bandwidth in the recording.
Capstan servo   A servo system that maintains proper capstan motor rotational speed during recording by comparing the phase of the capstan motor PG signal with that of an external signal from the video input terminal or with the internal sync signal output when there is no input from outside. During playback, proper rotational speed of the capstan motor is maintained by comparing the CTL signal phase with that of an external input or the internally generated sync signal.
Carrier leak   In a home VCR, the luminance signal is frequency modulated onto a carrier. If some of the unmodulated carrier appears during playback, a beat effect will appear on the screen as a series of wavy lines. If unmodulated carrier gets through in addition to the modulated carrier, the result is called *carrier leak*.
Chroma   The color portion of a video signal.
Chroma (degree of saturation)   When talking about any particular color, such as red, one may speak of dark reds and light reds. A dark red is a saturated color, but as white light is added, it will become less and less saturated and will be seen as a lighter red. Chroma refers to the degree of saturation, regardless of hue.
Chrominance signal   Both hue and chroma are combined in the chrominance signal.
Clamp   The process of giving an ac signal a specific dc level.

Clogging   The gaps in the video heads can become clogged with dirt or by abrasion of the tape surface, resulting in noise and failure of the VCR to record and play back properly. Also, if the CTL head becomes clogged, the servo will cease to operate. These effects can be eliminated by cleaning the clogged heads.

C-mount   A standard type of screw-in lens mount used on 16-mm movie cameras. It is often found on video cameras as well.

Color bar   Within a color test signal, the color bar contains bars of color (from left to right: white, yellow, cyan, green, magenta, red, and blue) and peak black and white. It provides a means to gauge the luminance and chrominance of the picture.

Color burst   Also called the *color sync signal*, the color burst consists of 8 to 12 cycles of the chrominance subcarrier added at the end of the horizontal sync signal of the TV signal. This burst signal is exactly the same phase and frequency as the chrominance subcarrier being transmitted and becomes the reference for phase comparison during reception in generating the continuous carrier necessary at the time of deriving the color signal. Amplitude is used for chroma determination and phase is used for hue determination. In the NTSC color signal, the chrominance subcarrier is at 3.58 Mhz. However, in video recording systems the color carrier is shifted below the FM signal in the so-called color-under technique.

Color signal   The composite color signal consisting of the luminance signal, the chrominance subcarrier signal and the color sync signal. The luminance signal conveys the brightness. The chrominance subcarrier conveys hue and saturation. The color sync signal is the color burst signal.

Color temperature   When heated to a high temperature, a black body emits light. The relationship between the color of the light and the temperature is constant, allowing color to be expressed in terms of absolute temperature. Reddish colors have a lower color temperature; bluish colors have a higher color temperature.

Compatibility   If a tape made on one VCR can be played back on another and vice versa, the tape is said to be compatible with both units.

Composite sync signal   A sync signal composed of the horizontal sync, vertical sync, and equalizing pulse signals. The composite video signal is produced by adding this to the video signal. If the signal is color, the color burst is included in the composite sync signal.

Composite video signal   The combination of video signal, blanking signal, and the composite sync signal.

Condensation (dew)   Because of the condensation of moisture in the air, water droplets will form on a cold piece of glass or metal in a warm room. This same kind of condensation will occur on the video head drum under certain conditions. If a VCR is used when there is condensation on the drum, the tape will stick to the drum and be ruined. Hence, a dew detector is provided in most home VCRs, and when dew is formed on the drum, the recorder is automatically shut off.

Control (CTL) signal   A control signal is necessary if the servo circuit is to maintain proper head rotational speed. During recording, the vertical sync signal is separated from the video input signal and recorded on a separate track (CTL track). This signal is picked up on playback and becomes the reference for servo system operation.

Control track (signal)   A signal recorded on the video tape that is to be used during playback as a reference for the servo circuits.

Converted subcarrier   In the so-called color-under technique, the color 3.58-Mhz subcarrier and its sidebands are shifted down to about 629 kHz in VHS systems and to about 688 kHz in Beta systems.

Crosstalk   The name given to unwanted signals obtained when a video head picks up information from an adjacent track on the video tape.

CTL head   Records and plays back the video control (CTL) signal. The CTL head and the audio head are generally in a single unit.

Cue and review   Forward and reverse viewing at a faster than normal speed with or without sound; originally called "Betascan" because it was introduced by Sony in their Beta format recorders.

DDC (Direct Drive Cylinder)   In VHS recorders it is common practice to drive the video head drum or disk by a self-contained brushless dc motor using no belts or gears. The drum and drive motor are inside a cylinder. Hence, the term *direct drive cylinder* is applied.

Dark clip   After pre-emphasis the negative-going spikes of a video signal may be too large in amplitude for safe FM modulation. A dark clip circuit is used to cut off these spikes at an adjustable level.

Delta factor (Δf)   A term used to indicate that a playback signal has some jitter—that is, "wow or flutter" or "a change in frequency." It means that the color signal off the tape is not stable at the down-converted color subcarrier frequency but is some small amount above or below that frequency.

Deviation   A term used to describe how far the FM carrier swings when it is modulated. In VHS recorders, the upper limit is 4.4 Mhz, and in Beta recorders, 4.8 Mhz.

Dew detector   A variable resistor, whose resistance value depends upon the ambient humidity, detects the presence of moisture on the video head drum.

Dihedral angle   A term used to describe the relative angular position between the video heads as they are mounted on the head cylinder. Perfect dihedral adjustment is realized when the tips of the heads are exactly 180 deg apart in the two-head case.

Disk   The rotating drum carrying the video heads is sometimes referred to as the *disk*.

Disk tach pulse   The signal used in the drum (disk) servo circuit to control the rotation speed of the video heads. Also referred to as the PG (pulse generator) signal.

Dropout   A momentary absence of FM or color signal off the tape caused by uneven oxide or a coating of dust or dirt on the tape or video heads.

Drum servo   The servo system which governs head rotation by comparing the phase of the vertical sync signal derived from the input video signal with that of the 30 PG signal produced by drum rotation. Any difference between the two is detected; as an error voltage, it is amplified and applied to the brake coil.

Duty cycle   Referring to a rectangular waveform, the duty cycle refers to the percentage of "On" time for one complete cycle. A waveform with equal periods of "On" time and "Off" time for one cycle is a square wave.

Editing   The process of rearranging, adding, and removing sections of the picture and sound already recorded on the tapes. Two different modes of electronic editing are used with video tape. One is called the *inset* mode and the other is called the *add* mode. The former means putting new picture and sound in the midst of a recorded section, whereas the latter is a technique of adding on different picture and sound. This mode is also known as the *assemble* mode.

E-E (Electronics-to-Electronics)   This is the picture viewed on the TV set when a recording is being made. It goes through only some of the electronic circuits and none of the magnetic components. It is called the *E-E mode*.

Emphasis   The process of boosting the level of the high-frequency portions of the video signal.

Encoder   A device that converts the color-camera video signal into an NTSC signal.

End sensor   An auto-stop device to detect the end of the recording tape. In VHS systems clear plastic leaders and phototransistors are used to provide this detection. In Betamax systems opaque metal leaders lower the Q of two oscillator coils at the end of the tape.

Equalizing pulses   A series of six pulses before and after the vertical sync signal to ensure proper interlaced scanning.

External sync   Besides the video signal produced by a camera tube, vertical and horizontal sync signals are necessary to make up a picture. These sync signals may be fed to the camera from outside or generated by an oscillator in the camera. The former is called *external sync*, the latter *internal sync*.

FG (Frequency Generator)   Pickup in Capstan base generates a signal used in phase comparison with a reference signal to develop capstan servo control.

Field   One half of a television picture. A field consists of 262.5 horizontal scanning lines across the picture tube. Two fields are displayed on the TV picture tube to form a complete picture, which is called a *frame*. After scanning 262.5 lines from top to bottom of the picture tube, the beam returns to the top of the tube to scan another 262.5 lines from top to bottom. The lines of the second field lie in-between the lines of the first field. This intermingling of the two fields is called *interlaced scanning*.

FM recording   Direct recording of the video signal is well nigh impossible because its octave range is from dc to 4 MHz. Therefore, a carrier wave produced by an oscillator in the VCR is frequency-modulated in accordance with the amplitude of the video signal. The FM signal thus produced is recorded on the tape instead of the original video signal. FM recording has the advantages of permitting recording down to dc and allowing for the use of limiter circuits to eliminate level fluctuations.

FM signal   The luminance portion of the video signal is used to control the frequency of an astable multivibrator. The output of this multivibrator is a frequency-modulated signal swinging from 3.4 Mhz to 4.4 Mhz (plus sidebands) in the case of VHS and from 3.5 Mhz to 4.8 Mhz (plus sidebands) for Betamax.

Focal length   The distance from the optical center of a lens to the principal focus (the point where parallel rays of light focus).

Flagging   The term used to describe a TV set's inability to accept unstable playback pictures from a video tape recorder. All home VCRs have some degree of playback instability. A TV set with a long horizontal AFC time constant may not recover from the VCR's instability before the active picture is being scanned. A bending or flapping from side to side of the top inch or so of the picture, called *flagging*, may occur.

Frame   One complete TV picture consisting of two interlaced fields.

Frame-by-frame advance   A series of still frames advanced under user control.

Freeze-frame   A mode in which the picture is motionless on the screen.

Gate   A circuit that will deliver an output only when a specific combination of its inputs are present. Gates have analog and digital applications.

Guard band   The empty space between video tracks provided to prevent crosstalk. In long-play video-recording modes, there are no guard bands. Adjacent video tracks overlap, but a good picture is obtained by azimuth recording methods that eliminate the effects of the overlapped tracks.

HD (Horizontal Drive Pulse)   The horizontal drive sync pulse that, along with the vertical sync pulse, is supplied by the VCR for camera operation.

Head cylinder   A cylindrical piece of metal housing the video heads. The tips of the heads protrude slightly from the surface of the cylinder so that they can scan the tape as the cylinder spins (also referred to as the *head drum* or *disk*).

Head gap   The tiny space between the two pole pieces of the ferrite core used in the video heads. Its width ranges from less than 0.3 micron to 0.5 micron.

Head protrusion   For maximum tape-to-head contact, the heads protrude slightly from the drum. Thus, they actually penetrate the tape to a minor degree (called *head intrusion*). Tape stretch and head wear may result from excessive head protrusion.

Head switching   The action of turning off, during playback, the video head not in contact with the tape. A particular video head will be turned off 30 times per second.

Head switching pulse   The signal that is applied to the head amplifier to perform head switching.

Helical scanning   The type of video tape scanning used in VHS and Betamax recorders. As the heads rotate in the horizontal plane, the tape passes over the drum diagonally, resulting in diagonal tracks on the tape. The term "helical" applies because of the shape the tape path describes as it is wrapped around the cylindrical drum. Helical scanning is particularly adaptable to simulated slow motion, fast motion, and still playback.

High-speed scan   Cueing or reviewing at higher speeds than normal playback speeds (up to forty times normal speeds) to locate a particular segment on a tape.

Interchangeability   A term used to describe how well a particular VCR will play back a tape that has been recorded on another VCR of the same format.

Interlacing   The property of the scan lines of two television fields to lie in-between each other. This method is referred to as *interlaced scanning*. In the NTSC system there are 525 scanning lines, with 262.5 in one field and 262.5 in the interlaced field.

Interleaving   A term used to indicate that the harmonics of the chrominance signal lie in-between the harmonics of the luminance portion of the video signal as it is viewed on a spectrum analyzer. Thus, the color information of a video signal does not interfere with the luminance information, although it is broadcast at the same time.

Internal Sync   TV cameras equipped with a built-in oscillator that generates the sync signals are said to have *internal sync*.

Jitter   Instability in the playback signal caused by tape or head speed fluctuations. The picture appears to have a rapid shaking program.

Loading   In cassette operation, an automatic loading mechanism is needed to take the tape out of the cassette and wrap it around the head drum and capstan so that recording and playback may take place. The term "loading" refers to the entire mechanical operation from cassette insertion to completion of tape path setup. "Unloading" refers to the opposite operation during which the tape is put back into the cassette.

Luminance   That portion of the video signal that contains the black and white information and the sync. In effect, the luminance signal conveys the information on picture brightness.

M-loading   A loading system used in VHS recorders in which the tape passes around the head drum in an M-shaped path.

MM (Monostable Multivibrator)   Usually an integrated circuit that gives a logic high or low output with a variable duration upon receipt of an input pulse or transition.

ND filter   An ND (neutral density) filter is an optical filter used on cameras. It passes all colors (light frequencies) equally with no color correction. The number following the letters ND indicates the fraction of the light passed by the filter. Thus, an ND4 passes one-fourth of the light, and an ND6 filter passes one-sixth of the light. Such a filter is used instead of a small aperture in bright conditions to shorten the depth of the field.

Noise canceler circuit   The video signal band includes harmonics of the sync and luminance signals, which cause cross color noise and must be removed to obtain good picture quality. Noise canceller circuits achieve this result by separating the signal and noise components with a high pass filter and a limiter and then applying the noise component with equal amplitude and reverse phase back onto the signal. Cancellation of the noise results.

Nonlinear emphasis   Similar to regular emphasis except that lower-level high-frequency portions of a video signal are given more boost than higher-level high-frequency portions.

NTSC (National Television Systems Committee)   These four letters identify the NTSC system of color TV used in the U.S.A., Canada, and Japan.

Overlap   The recorded video signal is reproduced by the alternating output of two video heads, but the beginning of head B output is allowed to overlap a bit with the end of the head A output in order to eliminate blank areas. The amount of this overlap is between three and twelve lines.

Overmodulation   When a large luminance signal input increases the oscillator frequency in the FM modulation circuit beyond the normal limits, overmodulation occurs. Its effects may be seen as a black edgelike effect over the right side of white portions of the picture.

Ω *(Omega) wrap*   When looking straight down on the drum of a helical scan VCR, the tape is seen to be wrapped about 180 deg around the drum. The name is derived from the shape of the tape wrap.

Pause   In this mode, the tape transport is stopped while remaining in the playback or record mode and differs from the stop mode in that one can go from pause into the record mode directly without having to press the record button again. The pause mode is useful during editing and for eliminating undesired sections of a program during recording.

PG (pulse generator)   The output of this generator is the head position signal. Also called the *disk tach signal*, it is used in the drum (disk) servo circuit to control the rotational speed of the video heads.

PG bipolar   Pulse generator signals that have both positive and negative excursions.

Photoconductive layer   The light passing through the lens in a TV camera is focused to form an image on the photoconductive layer on the target with the camera. Resistance varies as light strikes the target, and the resulting signal corresponds to the level of light at each point in the image.

Q   Designates transistors on schematics and parts lists.

Rewind   The tape transport mode in which the tape is wound at high speed onto the supply reel. Video tapes are played on only one side in VHS and Beta systems (in contrast with audio cassettes). Therefore, they must be rewound to be viewed over again from the beginning.

Rotary chroma   The process used in VHS recorders to change the phase of the chrominance signal at a rate of 15,734 times per second. Phase rotation of 90 deg between the video head signals is a part of the VHS azimuth method of eliminating the effects of track overlap.

Rotary transformer   The transformer that provides the video signal current to the rotating video heads during recording and picks up the signal from the heads during playback.

Rotating drum (disk)   The drum or disk that carries the video heads past the tape at a high speed. In helical scan VCRs, the tape is wrapped around the rotating drum diagonally (forming a helix) so that the rotating video heads can trace the tape surface at the correct diagonal angle.

Rotating head   For video recording, the speed of the head relative to the tape must be at least 200 times the normal audio tape speed. This speed is achieved by rotating the heads while maintaining a slow tape speed. In a helical scan system, the heads rotate in a horizontal plane as the tape passes around the drum diagonally. The tracks traced on the tape are therefore also diagonal. Compared to audio recording, this video recording technique greatly increases the recording area in terms of efficient use of the tape surface.

Sample and hold (S/H)   A process used in comparator circuits by which the value of a particular signal is measured at a specific moment in time and then stored for later use.

Scanner   Another name for the video head assembly which includes the heads, head cylinder, cylinder base, and head motor.

Separate mesh   In some vidicon camera tubes, a wire mesh electrode is employed in addition to the focusing electrode. The voltage of the former is 1.6 to 1.7 times that of the latter. Separate mesh vidicons have improved resolution.

Servo   A servo mechanism is a type of automatic governing system used to maintain a fixed speed, position, or angle of operation.

Signal splitter   A device that distributes a single signal equally to two or more other lines. Signal strength drops in proportion to the number of lines to which the signal is distributed (distribution loss). For instance, there is a 4-dB loss for a two-way distributor and a 7- to 8-dB loss for a four-way distributor. When distributor terminals are not being used, they should be connected to a dummy load.

Skew   The effect produced by tension error. It is the change of size or shape of the video tracks on the tape from the time of recording to the time of playback. Skew can occur as a result of poor tension regulation by the VCR or by ambient conditions affecting the tape.

Slow motion   Viewing at a speed that is a fraction of normal speed.

S/N ratio   The ratio between the desired signal and the unwanted noise components. The larger the S/N ratio, the clearer the picture and sound.

Standard lens   The size of the camera tube determines the standard lens for a video camera. For example, the focal length is 25 mm for a 1-in. vidicon and 16 mm for a standard of a ⅔-in. vidicon. A lens longer than the standard focal length for a certain size of camera tube is called a telephoto lens. If it is shorter than standard, it is called a wide-angle lens. A zoom lens offers continuously variable focal length from wide-angle to telephoto without affecting focus or aperture.

Subcarrier   The name of the 3.58-MHz continuous wave signal used to generate the quadrature chrominance signal. In the VHS and Betamax systems, the 3.58-Mhz signal is down-converted to 629 MHz and 688 MHz, respectively, to place the color signal under the FM modulated luminance signal frequencywise.

Sync generator   An internal or external generator supplying the sync signal required by a video camera.

Sync signal   The pulse signals used to synchronize the operation of video equipment. The horizontal sync signal determines the start of horizontal scanning; the vertical sync signal determines the timing of the beginning of vertical scanning.

Tape Counter   A counter on the VCR that advances as the tape is used. Its figures may be used for finding a specific place on a tape or judging the amount of tape remaining.

Tape guide   Guides to ensure that the tape properly passes the video heads, the CTL head, the capstan, and other areas along the tape path. They usually consist of metal cylinders or posts, some of which may be tapered.

Tape path   The complete path taken by the tape as it leaves the supply reel; goes by the tape guides, the erase head, head drum, and CTL head; is driven by the capstan and pinch roller; and wound onto the take-up reel.

Tape pattern   The magnetic pattern on the tape composed of the video tracks, control track, and audio track.

Target   That part of the vidicon camera in which the image is formed.

Tension error   See "Skew."

Tension servo   A servo system that automatically applies a brake to the supply reel as needed in order to maintain constant tension along the tape between the supply and takeup reel spindles.

Test pattern   A pattern used to check transmission and equipment performance. Test patterns are designed to check resolution, contrast, aspect ratio, the presence or absence of scanning distortion, and other picture characteristics at a glance.

Time base stability   A term used to describe how closely the playback video signal from a VCR matches an external reference video signal, with regard to sync timing rather than picture content.

Timer recording mechanism   A system that automatically triggers a VCR or audio tape recorder to begin recording (or playback) at a preset time.

Tracking   The action of the spinning video heads during playback when they accurately track across the video RF information laid down during

recording. Good tracking indicates that the heads are positioning themselves correctly and are picking up a strong RF signal. Poor tracking
indicates that the heads are off track and picking up a low level RF signal
or noise.

Tracking adjustment   When a tape is recorded by a VCR having a CTL
head out of position, is played back on another unit, the video heads may
not trace the tape at exactly the correct point and mistracking will result.
This tracking error can be corrected electronically by delaying the CTL
signal rather than actually moving the CTL head. Such an electronic
control is called *tracking adjustment.*

Trinicon   A color camera tube developed by Sony. The Trinicon uses
one camera tube for picking up the three primary colors of red, green, and
blue.

U-loading   A loading system, used in Betamax recorders, in which the
tape passes around the head drum in a U-shaped path.

Unloading   In contrast with loading, which refers to the automatic
wrapping of the tape around the head drum, *unloading* refers to returning
the tape to its original position in the cassette after recording or playback.

VCO (Voltage Controlled Oscillator)   An oscillator whose frequency of
oscillation is controlled by an external voltage.

VHS   Video Home System

Video head   The ferrite electromagnet used to develop magnetic flux,
which will put RF information on the tape. The same heads record the
video signal on the tape and pick up the signal during playback.

Video track   The name given to the RF information laid down during
the recording, as a particular video head scans across the tape.

V-V (Video-to-Video)   The actual picture produced from a tape during
playback.

Vidicon   A type of camera tube whose role is to convert the optical
image formed by the lens into an electronic signal.

Viewfinder   On a video camera, the viewfinder can be either optical or
electronic. If electronic, it is a small TV monitor, with a small diameter CR
tube, that shows the picture shot by the camera lens. It is placed in such a
way that it gives the operator the impression of actually looking through
the lens so that he can easily adjust focus and aperture and follow
movement of the subject.

VXO (Voltage controlled crystal oscillator)   Similar to a VCO except
that a quartz crystal is used as a reference. In this case, the crystal
oscillator frequency can be varied slightly.

White balance setting   When shooting with a color camera under
varying light conditions, it is necessary to decide on a color reference on
which the expression of other colors can be based. White balance concerns
the process of establishing this balance. This requirement results from the
fact that an object will appear to change color depending upon the color
temperature of the light illuminating it.

White clip   After emphasis, the positive-going spikes (overshoot) of the video signal may be large enough to result in overmodulation of the FM signal. A white-clip circuit is used to cut off these spikes at an adjustable level.

Wow and flutter   Tape transport fluctuations that may cause a regularly occurring instability in the picture and a quivering or wavering effect in the sound during recording and playback. Lower frequency fluctuations (below 3 Hz) are called *wow*; higher frequency fluctuations (3 to 20 Hz) are called *flutter*.

Y signal   The black-and-white portion of a video signal containing the luminance and sync information.

Zoom lens   A lens with a continuously variable focal length (*see* "Standard Lens").

Index